AF552614

Management of Wastelands

Volume-II

Problems and Prospects

Volume. I : Identification and Distribution

Volume. II : Problems and Prospects

Volume. III : Reclamation and Development

About the Editor

Hridai R. Yadav, an Eminent Scientist obtained degrees of M.Phil., Ph.D. from Jawahar Lal Nehru University, New Delhi, worked with CSIR Government of India as Research Scientist and Pool Scientist for eight years having authored 6 books and 63 research papers. He has been the Member of National Wastelands Development Board, Government of India and nominated as UGC observer for the selection of Professor in the various Universities and UGC nominee in the Board of Management of Autonomous Institute of Engineering and Technology (GBTU), Lucknow. H.R. Yadav has been the Member of various Government and Autonomous Institutions/Universities, besides being Founder-Member-Secretary-Director of the Institute of Wastelands Reclamation and Rural Development (1990), Institute of Science and Technology for Rural Development (1994) and UGC accredited RIWARD Post Graduate College, Hanumanganj, Sultanpur, U.P. (2004). Presently, H.R. Yadav is engaged in the implementation of Agro-Afforestation on Patta Banjar Land alloted to Scheduled Caste Community, Women Empowerment and Rural Health Mission Programmes in Amethi.

Management of Wastelands

Volume-II

Problems and Prospects

Edited by

Hridai R. Yadav

CONCEPT PUBLISHING COMPANY PVT. LTD.
NEW DELHI-110059

ISBN-13 : 978-81-8069-943-6 (Set)
978-81-8069-944-3 (Vol. I)
978-81-8069-945-0 (Vol. II)
978-81-8069-946-7 (Vol. III)

First Published 2013

Published and Printed by

Concept Publishing Company Pvt. Ltd.
A/15-16, Commercial Block, Mohan Garden
New Delhi-110059 (India)
T: 25351460, 25351794, *F*: 091-11-25357109
E: publishing@conceptpub.com, W: www.conceptpub.com
Editorial Office: H-13, Bali Nagar, New Delhi-110 015, India.

Cataloging in Publication Data--*Courtesy:* D.K. Agencies (P) Ltd. <docinfo@dkagencies.com>

Management of wastelands / edited by Hridai R. Yadav.
3 v. cm.
Includes bibliographical references and indexes.
Contents: v. 1. Identification and distribution — v. 2. Problems and prospects— v.3. Reclamation and development.
ISBN 9788180699436 (set)
ISBN 9788180699443 (v. 1)
ISBN 9788180699450 (v. 2)
ISBN 9788180699467 (v. 3)

1. Waste lands—India—Management. I. Yadav, Hridai Ram, 1957-

DDC 333.730954 23

Dedicated

To

My Publishing Mentor

Late Shri Naurang Rai Ji

Preface

It has been realized that ever increasing demand for Fuel Wood, Fodder, Fiber, Fruits, Fisheries and Food Grains (F^6), of the growing population and increasing needs of the raw materials for the Industries has led to environmental degradation and rapid expansion of wastelands. It has been noted that out of the total land mass of 328 million hectares of our country, 175 million hectares land are degraded. The land degradation is mostly caused by the poverty and mismanagement of land. The degradation of natural land resources proceeds from hacking, overgrazing, excessive agricultural cropping, overirrigation and excess use of fertilizers, mismanagement and natural calamities etc. has caused land degradation. The Satellite imagery has indicated that we are loosing 1.3 million hectares of forest every year. The fire wood requirement is 130 million tonnes per year out of which we are fetching about 50 million tonnes per year from forest and 80 million tonnes balance fire wood requirements has yet to be generated. The larger revenue raised through unplanned and unscientific cutting of forest, the greater destruction of forest area and it has to be stopped.

Article 48-A of the Constitution of India enunciates, "The State shall endeavour to protect and improve the environment and to safeguard the forest and wildlife of the country". For the fulfilment of this objective, the task of environmental protection has become of prime importance to the nation. The schemes of environmental protection, afforestation, prevention and reclamation of wastelands, better use of irrigation through scientific management and development of alternative sources of rural energy have already been included in the 20-point economic programme.

The major themes for discussion are : (i) identification, definition, classification and distribution of wastelands, (ii) reclamation of wastelands for Agro-afforestation, (iii) water conservation and irrigation management, (iv) environment protection, (v) economics of wastelands development, and (vi) people's participation and generation of awareness, policy planning strategies.

Wastelands problem is mostly man-made and causes misery to millions of the rural poor. Thus reclamation of wastelands may be adopted as a strategy for the extension of net sown area according to its suitability to increase the total agricultural production through cropping or for the plantation programme predominantly for the fulewood and fodder for the overall development of the rural poor. It is felt that there is an immense need : (i) for a comprehensive survey, mapping and identification of wastelands and preparation of an action plan for wastelands reclamation and to create the data base, (ii) to save existing forests by enacting strict laws, (iii) to increase and maintain forest area for industrial use through proper management, (iv) to enhance wood wealth and to meet fodder needs through social forestry. It should also include fruit trees. A coordinated action plan should be drawn for social forestry and Agro-forestry. We must also examine the role and impact of social forestry and the status of actual forests. We cannot allow actual forest to be destroyed and expect social forestry to compensate. It would help to maintain the ecological balance and protection of environment for its longer sustained use by the resourceless village poors.

The importance of the wastelands and land degradation has been stressed by many environmental scientists, ecologists and social scientists. In totality our environmental policy, the proper management of natural resources must take absolute primacy. Management, enrichment and optimum utilization of scarce natural resources are vital prerequisites of a purposeful national strategy for economic development and poverty alleviation.

It was the vision and environmental zeal of Indiraji, that led to her pioneering statement on Environment at the Stockholm Conference. With the subsequent establishment of the Department of Environment, we have gradually come to realize that, in addition to control of industrial pollution and protection of wildlife, we must prevent farther degradation of the country's basic natural resources and endowments, and life support systems of land, water and vegetation. The setting up of the National Wastelands Development Board and

Department of Wastelands Development and later on Department of Land Resources has been formed in the Govt. of India yet another important milestone in the saga of our efforts to prevent environmental degradation and create tree cover over millions, of hectares land, in the process, to promote the economic and social well-being of millions of our people.

The preservation and protection of our environment and forests, cannot secured by statutory measures, governmental control and punitive mechanism. The realization that the preservation and enrichment of our environment and the propagation of our forest cover is an ecological imperative must become ingrained in the national psyche. We must start with the children, in their schools and colleges The awareness that we create now in these young minds that the Tree is India's life and part of it would be an awareness that will reach out into generations in the future as well. In the historic words of Indiraji :

> "In his arrogance with his own increasing knowledge and ability, man has ignored his dependence on the earth and has lost his communion with it. He no longer puts his ear to the ground so that the earth can whisper its secrets to him. The national song which inspired our freedom movement describes our land as one endowed with water and fruit, rich with the greenness of growing plants. We must make this true not only of India but of all lands."

If her dream is to come true, we must no longer confine ourselves to regulations and control to preserve and enrich our natural resources, but each one of us must put his ear to the ground and listen to what Mother Earth asks of him or her. And having listened, each one must do his or her bit for giving back to Mother Earth what centuries of despoliation has taken away so ruthlessly from her.

In view of the above we must respond to generate proper movement for conservation of available wastelands natural resources. Decentralized village peoples nurseries and tissue culture raised plants or hybrid plants, and reclamation of wastelands, plantation of Agro-forestry and social-forestry species and leasing of the land to the interested rural poor on Tree Patta basis for Agro-Afforestation and a comprehensive awareness raising programme should be taken up under the MNREGA Scheme through village Panchayat and the State - Districts - Tehsil - Block and Village Panchayat level Administrative machinery must create an infrastructure and atmosphere conducive to MNREGA Job Card holders to generate employment and to ensure people involvement in the Agri.-Horti.-Forestry programmes at grass root level.

Village Panchayat level *Van Mahotsava* must be re-initiated in corporating the reclamation of wastelands for Agri.-Horti.-Afforestation programmes with the view to meet the increasing demand of growing cattle and human for Fuel Wood, Fodder, Fiber, Fruits, Fisheries and Food Grains (F^6) and to generate employment and increase the income of the deprived resource less rural poor peoples.

In view of the above an attempt has been made to incorporate the various case studies carried out by the agricultural scientists, environmentalists, social workers and policy planners at various levels with the view to provide exchange of views, opinions, experiences for further formulation of peoples participatory planning process for Agri.-Horti.-Afforestation programme through reclamation of wasteland natural resources to meet the ever increasing demand of growing cattle and human population and also to maintain the ecological imbalances at the grass root level.

It has been realized that the various programme related to soil conservation, wastelands development, environment protection and management, Agri.-Horti.-Afforestation management does not normally percolate the basic strata of the common people living in the remote rural areas has often led to its failure at the grass root level. Hence, a socio-politico, techno-scientific peoples participatory wastelands reclamation for the Agriculture, Horticulture and Social-forestry and environmental management programmes must be initiated at grass root level through peoples participation and the usufructs accrued out of it should be shared by the resourceless deprived rural poor multitudes at village level. The administrative machinery has to be properly involved and the Mahatma Gandhi National Employment Guarantee Act (MNREGA) Scheme should generate 100 days employment for the MNREGA Job Card holders for Reclamation and Conservation of the allotted Patta Banjar land for Fuel Wood, Fodder,

Fibre, Fruits, Fisheries and Food Grains (F^6) to meet the ever increasing demand of growing cattle and human population and to improve the degraded environment of the area and also to improve the socio-economic conditions of the deprived rural poor multitudes living in the remote rural areas.

Prof. (Dr.) Hridai R. Yadav

[illegible] [illegible] to meet the [illegible] increasing demand of growing cattle and human population and to improve the degraded environment [illegible] and also to improve the socio-economic condition of the deprived [illegible] poor [illegible] in the [illegible] rural areas.

Prof. (Dr.) [illegible] R. [illegible]

Acknowledgements

It is known fact that from time immemorial, man has lived in harmony with nature, in a symbiotic relationship with the nature. The nature gave its bounty to man was in return, recompensed adequately by self-regenerating process or eco-development, built into the cultural, social and economic traditions of human life, as an individual and in groups.

In view of the above, an attempt has been made to re-capsulate the techno-scientific practical studies carried out on Wastelands identification, distribution and Problems and Prospects of Wastelands and Reclamation and Development of Wastelands, which had been printed in the earlier work, with the view to provide decadal changes in the distribution, problems, prospects, reclamation and policy, planning strategies of the wastelands development at national and grass root level.

We are highly obliged and grateful to all the contributors for their already contributed techno-scientific studies which has been re-capsulated to bring out new volumes of Wastelands for its wider replicability. We acknowledge our sincere gratitudes for all the contributors for their kind support, without their help this re-capsulated work on Wastelands would not have been possible.

We acknowledge our sincere gratitudes to Late Sri Naurang Rai Ji who supported in an integrated manner for publication of Wastelands studies carried by me before to decades.

We are grateful to Sri Ashok Kumar Mittal, CMD, Concept Publishing Co. Pvt. Ltd., New Delhi, to kindly accept the publication of re-capsulated contributions on Wastelands in the Loving Memory of my publishing Mentor Late Sri Naurang Rai Ji.

Last but not the least, I am grateful to my wife Reeta and lovely son Sahit Hridai, without whose support completion of this work would not have been possible.

Prof. (Dr.) Hridai R. Yadav

Introduction

An effort has been made in this Volume to elaborate the problems and prospects of wastelands. The causative factors responsible for the growth of wastelands have been explained by the various Scientists and the various reclamation procedures adopted for wastelands development for Agriculture, Horticulture, Social-Forestry programmes to meet the ever-increasing demand of growing cattle and human population for Fuel Wood, Fodder, Fibre, Fruits, Fisheries and Food Grains (F^6).

The Reclamation Suitability Classification of Wastelands has been explained by H.R. Yadav and he has also attempted to analyze the Agro-Afforestration Management on Wastelands. The factors responsible for the growth of Wastelands has been elaborated in detail by H.R. Yadav. While R.P. Dhir and A.S. Kolarkar have attempted to explain the Salt-affected Wasteland Soils of Arid Zone and their Development Potential. D. P. S. Verma has made an attempt to explain the Wastelands Development with a Case for *Prosopis Juliflora*. V.N.K. Pillai has explained the Wasteland Development in Haryana. S.S. Verma has explained the Impact of Irrigation on Wastelands : A Case Study of Gandak Command Area (U.P.). A.C. Chaturvedi has explained Structural Measures for Wastelands Development and D.B. Misra, S.L. Dabral and A.K. Gulati have explained the Evaluation of NRSA Wastelands Map from User's Angle : A Case Study in Meerut, Ghaziabad and Bijnor District of Uttar Pradesh. C.B.S. Dutt, P.P. Nageswara Rao, B. Manikiam, A.K. Gupta, G. Behara and K. Ganesha Raj have attempted to explain the Identification and Mapping of Wastelands Using Remote Sensing Technique : An Approach at the Micro level. N.C. Gautam, L.R.A. Narayan, G.Ch. Chennaiah, V. Raghav Swami and R. Nagraja have attempted to explain the Use of Remote Sensing Techniques for Mapping and Monitoring of Wastelands in India. J.S. Yadav has made an effort to explain Wasteland

Development of Mewat Region of Haryana, its Identification and Planning at grass root level. P.D. Bajpai has explained the Wasteland Reclamation and Development in Uttar Pradesh. K.K. Mehta has attempted to analyse the New Approach for Growing Eucalyptus trees on Alkali Soils. S.S.S. Shastri has made an effort to explain the Levelling and Afforetation of one Milliion Hectare Dotty Land situated along the banks of Chambal river.

I.S. Singh and R.K. Singh have made an effort to explain the Fruits as a Component of Agroforestry. R.K. Pathak, S.N. Dixit and R. Dwivedi have tried to explain the Prospects of Aonla, Ber and Bael Cultivation on Salt-affected Wastelands in India. K.K. Mehta has explained the Experiences of Growing Crops and Trees on Salt-affected Soils of Hardoi District in Uttar Pradesh. A.K. Saxena, P.K. Singh and B.P. Singh have made an attempt to explain the Effect of Mango (Mangifera Indica L.) Trees on Growth and Yield of Wheat (Triticum Aestivum L.). M.R. Verma and P.M. Singh have tried to explain the Scope of Power Tiller for Agroforestry Management.

Prof. (Dr.) Hridai R. Yadav

Contents

List of Tables

List of Figures

1

Reclamation Suitability Classification of the Wastelands

Hridai R. Yadav

Introduction

To make a sound and conversion plan one should establish the suitability classification of wastelands for specified uses. The suitability classification of wastelands is a systematic arrangement of different kinds of land according to those properties that determine the ability of land to produce on a virtually permanent basis. Suitability for cultivation is assumed to include the use of machinery at least of plough, tillage implements, harvesting equipment and the capacity for at least a moderate yield of one or more crops with suitable treatment and protective measures. The restrictions imposed by natural land characteristics necessarily affect (a) the number and complexity of the corrective practices to be used, (b) the productivity of the land and (c) the intensity and manner of land use. The classification is made and used for practical purposes, which is the selection and application of wasteland uses and treatments that will, while using the wastelands, keep it in condition for long-time production. The latter will involve the control of erosion, conversion of rainfall and maintenance, the classification is arranged according to a number of categories. The first category covers the land suited for cultivation and not for cultivation, the second category covers eight land capability classes. The classes under this category are differentiated according to the degree of limitation in land use, imposed by best and most easily formed land (Class I) to land which has no value for cultivation, grazing or forestry but may be suitable for wildlife, recreation and watershed protection. Thus, the capability is the inherent quality of lands to perform at a given level of general agricultural use while suitability is a statement of adaptability of a given area for a specific land use type. On the other hand, wastelands capability classification is a method of grouping wastelands on the basis of their classes assessed in relation to their suitability to agricultural uses envisaged in the plan. It relates to the morphometric, climatic, hydrologic, pedogenic and other similar environmental conditions affecting the agricultural use and productivity.

The wastelands suitability classification will help to organise significant factors for conversion. It is an imperative grouping and grading of soils, according to their potentialities and limitations, their capacity of producing agricultural crops and responsiveness to management practices. This is a system to make a sound and complex form of conservation plan to find out the suitability of wastelands for specific areas and its agricultural uses.

Thus, the methods of wastelands suitability classification is a systematic segregation of different kinds of wastelands which are distinguished from one another by variation in the kind and degree of use imposed by soil characteristics morphometric, hydrologic, climatic and pedogenic, including other environmental factors. Thus, the fundamental purpose of wastelands suitability classification is to utilise the wasteland resources according to their capabilities through rational planning. The suitability classification of wastelands are mainly based on the environmental parameters including inherent soil characteristics and field observations. The following five suitability classes have been ascribed to the various wastelands :

Suitability Class I—Very easily Reclaimable
Suitability Class II—Easily Reclaimable
Suitability Class III—Reclaimable with certain difficulty
Suitability Class IV—Reclaimable with moderate difficulty
Suitability Class V—Reclaimable with great difficulty

Very Easily Reclaimable

Those wasteland types are very easily reclaimable where the

environmental severity is very low and man with his low level of technology is able to interact for availing wasteland resources for agricultural use. In this category other types of wasteland and fallow land (1-2 year) are classified because these two types of wasteland have less severe natural constriants and there was time when these wasteland types were under agricultural use but due to deteriorating soil conditions are presently left out of cultivation.

Easily Reclaimable

Old fallow land, Banjar, Waterlogging and Riverine wastelands are classified as easily reclaimable under the wastelands suitability classification. Such wasteland types were previously under cultivation but presently are left out of cultivation due to deteriorating soil fertility status, erosion and accumulation of the water accordingly. The man-made problems as well as natural constraints of such wasteland types may be improved through good deal of agricultural technological inputs and through the managed suitable agricultural practices. The problems of such wasteland types have been mainly caused, by overcropping, large size of landholdings, mismanagement and carelessness of the farmers, deteriorating fertility status of the soil. Such wastelands can be easily reclaimed through proper attention and agricultural technological doses. These wasteland types are also recommended for agricultural uses and for producing crops according to its suitability classifications.

Suitability Classification of the Wastelands

Suitability Class I	*Suitability Class II*	*Suitability Class III*	*Suitability Class IV*	*Suitability Class V*
Very easily Reclaimable	Easily Reclairnable	Reclaimable with certain difficulty	Reclaimable with moderate difficulty	Reclaimable with great difficulty
Fallow and other types of wasteland	Old fallow Banjar land, Waterlogging & Riverine	Usar, Gullies or Ravinous and Shifting cultivation area	Sands, coastal sand dunes, Degraded forest land	Undulating land, Mining and Industrial Wasteland, and Kankrili land

Reclaimable with Certain difficulty

Those wasteland types, where environment is severe to a certain extent and man with his proper level of technology may interact to avail such wasteland resources are classified in this category. Such wasteland types may be reclaimed with certain difficulty. Usar, Gullies or Ravinous and Area under shifting cultivation are classified in this category of suitability classification of wastelands. These wasteland types are used by the farmers but certain environmental constraints, mismanagement and carelessness of the farmers have created the problem in agricultural practices by which the farmers are bound to leave the land uncultivated.

Reclaimable with Moderate Difficulty

Wastelands which are reclaimable with moderate difficulty are those wasteland types where nature is more severe than the reclaimable with certain difficulty wasteland types, while man is leas permissible with his low level of technology to utilize such wasteland resources. The sands, coastal sand dunes and degraded forest land are classified under this category of wastelands suitability classification and these wasteland types can be reclaimed with moderate difficulty. It has been observed from the affected areas that these wasteland types have severity of environmental constraints and man is less permissible to interact with his low level of technology to avail these wasteland resources for agricultural purposes or for plantation programmes.

Reclaimable with Great Difficulty

Those wasteland types are classified as reclaimable with great difficulty where environmental factors are very severe, institutions restrictive or man with his low level of technology is not able to exploit these wasteland resources. Undulating land, Mining and Industrial wasteland and Kankrili land are classified in this category of wastelands suitability classification and such wasteland types can be reclaimed with great difficulty. The morphometric, climatic, hydrologic and pedogenic factors including other environmental factors are high enough not to permit the man for agricultural practices with his low level of agricultural technology. Thus, such wasteland types can be reclaimed with great difficulty with doses of agricultural technological inputs through proper management of agricultural practices.

Conclusion

The wastelands suitability classification is the best illustration of the useful work. It may be said that such suitability classification of wastelands is of immense importance and of practical utility in the wastelands management far agricultural development. Although it needs enormous labour and technical knowledge together with huge funds, it would help greatly in predicting the suitability of wastelands for various agricultural and plantation programmes. On the basis of suitability classification of wastelands, it has been found that the fallow and other types of wasteland can be very easily reclaimed, old fallow, Banjar, waterlogging and riverine land can be easily reclaimed, usar gullies or ravinous and area under shifting cultivation can be reclaimed with certain difficulty, while sands, coastal sand dunes and degraded forest land can be reclaimed with moderate difficulty and undulating land, mining and industrial wasteland and Kankrili land can be reclaimed with great difficulty.

REFERENCES

"A Manual on Conservation of soil and water", Handbook of Professional Agricultural Workers, USDA, Oxford Publishing Co., 1984.

District Sultanpur Land Revenue Records, Sultanpur.

Dhawan, C.L., "Land Reclamation", Central Board of Irrigation Power, Publication No. 43, Vol.III.

F.A.O., 1976, "A Framework for Land Evaluation", *Soil Bulletin* No.l.

Stamp, L.D. "The Land Utilization Survey of Britain", *Geographical Journal London*, Vol. 78, 1931.

Uppal, H.L., "Reclaiming Alkali Wastelands", *ICAR, Bulletin*, 85, 1961, New Delhi.

Yadav, H.R. "Genesis and Utilization of Wastelands", 1986, Concept Publishing Co. (P) Ltd, New Delhi.

2

Agro-Afforestation Management on Wastelands

Hridai R. Yadav

1. Introduction

The magnitude of the problem of Wastelands has been fairly well documented. The National Commission on Agriculture described in its report of 1972 that as much as 175 million hectares out of a total of 266 million hectares which are available for agricultural use are Wastelands, due to degradation of land of some form or the other. The explosive increase of the population, increasing desire of man to exploit marginal and sub-marginal lands for a modicum of return, unmindful of the further land degradation, man causes defective land use and cropping patterns and efficient water management have emerged as one of the main factors.

In particular, the loss of forest cover, has been alarming, as much as 40 million hectares forest land out of 74 million hectares is degraded forest land. In has also been estimated that our country is losing 1.5 million hectare of forests and 12,000 million tonnes of top soil every year due to deforestation and run-off. The process of rehabilitation may take anything from 500 to 1,000 years to restore one inch of top soil and up to 100 years to re-establish a good natural forest cover.

The relationship of man and nature, the symbiotic land between the rural poor and forests especially among Scheduled Castes, has been closest through the years. The traditional rights of such Scheduled Caste communities to minor forest produce, to grass and fallen dry wood for fuel have kept the rural communities going through the centuries. The man and nature relationship now stands threatened. Afforestation efforts, as an endeavour, is rendered more difficult since the needs of the community and those of the individual are at variance with each other and cannot be reconciled unless poverty amelioration programmes raise the level of living of those surviving below the poverty line. So far there has been a slackening of afforestation on the one hand and absence of support of the local communities for the protection and augmentation of forests on the other.

Agro-afforestation programme must be initiated as a people's movement. The agro-afforestation programme must be implemented by the people and for the people, through leasing of forest and non-forest land and wastelands to the deprived rural poor, the tree *patta* schemes must be reformulated and procedures should be simplified, tree growers co-operatives must be promoted and farmers should be encouraged to undertake tree plantation on their own farms and agricultural field boundaries. So far as the idea of leasing of wastelands is concerned, it has been observed at the field level that individual must have a vested interest in growing trees thereon. If our programmes on agro-forestry and social-forestry and under poverty alleviation schemes, confine themselves to providing wage employment to the rural poor, the landless and the unemployed, in ongoing agro-afforestation programmes and such programmes are foredoomed preciously because they fail to create such a vested interest among the people to plant trees, and to look after and maintain them with a view to enjoy the usufruct obtained from the agro-afforestation programmes.

2. Agro-afforestation Systems

Agro-afforestation is a system of land use along with combined growing of agricultural crops with social forestry – animal husbandries including food grain, fodder/fuel wood, fibre, fruits, fisheries and vegetables. Common cultivators has been engaged in cultivating agricultural crops along with social-forestry, and the domestication of livestock — wood trees, fuel wood – fodder – fibre – fruits and vegetables etc. to fulfil their day to day requirements, and the farmers never stopped using the trees, in response to agro-ecological conditions of the area, in about 700 B.C. man changed from hunting and food-gathering to agricultural crop-food production.

The promotion and development of agro-forestry—aims, potential and positive interaction between ecological and economic—agricultural and forestry activities — can be emphasized when proved scientifically. The increased production and productivity, sustainable land management and ecological conservation are some of the important objectives of the agro-afforestation system.

The restoration of ecological and environmental degradation and to improve the socio-economic and ecological crisis sustainable development of land resources have to be managed and maintained, the sustainable development comprises of the preservation of environment, equitable development, scientific technological change is a dire necessity for obtaining self-sufficiency in food grains, fodder, fuel wood, fibre, fruits, fisheries, vegetable and other produce. An unbroken symbiotic relationship with man and nature has to be re-established. Development should aim at improving the ecosystem and the environment for living cattle and human people providing them water, shelter, reclaiming the wastelands for greening and interior rural areas habitable. The higher standard of living and increased agro-afforestation productivity must be achieved with despoiling environment and nature of its beauty, freshness and purity of nature are so essential for the survival of downtrodden people surviving in the remote rural areas.

It has been observed that there has been a shift in the policy during the recent past from sustained productivity to sustained development. The ecological, socio-economic, scientific and technological inputs are important for determining sustainable management systems. In the present scenario, it is required to link the ecosystem processes with the multifaceted socio-economic activities centred for the sustainable development of a region.

There are different opinions and views regarding Agro-forestry systems like Agro-forestry is not a perfect system because there has been both success and failure with the various agro-forestry systems adopted. No doubt, agro-forestry has provided viable alternative of land use system with a suitable environment, and compatible social conditions, agro-forestry system should be helpful in providing substantial output. A good agro-forestry system should be able to solve ecological, land and environmental degradation crisis and socio-economic crisis and problems and it should be helpful in overcoming the physiological, ecological and environmental protection constraints that causes hindrance in the agroforestry development and adoption of the agro-forestry system for the sustainable development of the area.

Thus, there is tremendous pressure of the human population which has resulted in deforestation and degradation of land and degradation of valuable forest resources. In is now an established fact that unless there is stabilization of population participatory management for the reclamation of wastelands and agro-afforestation management for sustainable development to meet the increasing demand of growing cattle and human population for fuel wood, fodder, fibre, fruits, fisheries and food grains etc.

3. Afforestation in Uttar Pradesh

The cultural systems of our country has symbiotic relationship with the forest. Forests are the land's largest and most important ecosystem of the area. Forests have profound influence on the structure and functions of the human habitat, the forest is one of the prestigious property for the human being. Man is dependent on forest from pre-historical period. Explosive population growth and increasing demand of cattle and human population has caused deforestation and land degradation. Awareness for the afforestation is the need of the hour. As per the National Forest Policy, one-third land of the total geographical area must be afforested from good quality forest cover. The total geographical area of the Uttar Pradesh is 2,40,928 km. Forest Survey of India, Dheradun, Report-2003 indicated that there are only 14,118 sq. km. land which is afforested and it is only 5.86 per cent forest area to the total geographical area of the Uttar Pradesh.

Central Afforestation Council has been established in 1948, to strengthen the forest conservation and development programmes. *Van Mahotsava* programme had been launched during the year 1950 with the objective of the popularization of afforestation programme among the people. New Forest Policy of India was launched during the year 1952 with the objective of planning and implementation of different afforestation programmes. National Forest Policy was launched during 1988 to increase the afforestation area and the Government of Uttar Pradesh also implemented "Uttar Pradesh State Forest Policy, 1998".

(i) Highly Dense Forest

There are 1297 sq. km. area under highly dense forest in Uttar Pradesh. District Shrawasti has 210 sq. km. and Balrampur – 144 sq. km. highly dense forest, Bijnor District has 42 sq. km., Chandauli District– 2 sq. km and Gond District has only one sq. km. highly dense forest cover. The Kheri District is having 366 sq. km. Maharajganj District– 202 sq. km., Pilibhit 290 sq. km. Rampur 3 sq. km., Shahjahanpur – 20 sq. km. and Sonbhadra District is having 17 sq. km. highly dense forest in Uttar Pradesh. Thus, the highly dense forest area in Uttar Pradesh is very insignificant.

(ii) Dense Forest

Uttar Pradesh is having 4,699 sq. km. dense forest area. Dense forest area is found very high (846 sq. km.) in Sonbhadra District, Kheri (502 sq. km.), Mirzapur (316 sq. km.) and Chitrakut (346 sq. km.) in Uttar Pradesh. Dense forest area between 200 to 230 sq.km. is found in Pilibhit, Shrawasti, Balrampur, and Bijnor Districts of Uttar Pradesh. Dense forest between 100 to 200 sq. km. is recorded in Lalitpur, Maharajganj, Lucknow, Saharanpur, and Chandauli District, while dense forest between 50 to 100 sq. km. area is observed in Agara, Gonda, Hamirpur, Jalaun, and Shahjahanpur District of Uttar Pradesh. District Kannauj, Sant Kabir Nagar, Sant Ravidas Nagar, Ballia district are without dense forest and other districts are having less than 50 sq. km. dense forest area in Uttar Pradesh.

(iii) Open Forest

There are 8,122 sq. km. open forest area in Uttar Pradesh. Very high open forest area 1,606 sq. km. is found in Sonbhadra District, followed by Kheri – 446 sq. km., Lalitpur – 426 sq. km., Mirzapur – 466 sq. km., Shrawasti District – 347 sq. km., Chandauli – 327 sq. km. open forest area. District Agara is having 199 sq. km. open forest area, followed by Bijnor (129 sq. km.), Etawah (139 sq. km.), Hamirpur (111 sq. km.), Hardori (118 sq. km.), Jalaun (179 sq. km.), Jhansi (168 sq. km.), Lucknow (183 sq. km.), Maharajganj (118 sq. km.), Pilibhit (203 sq. km.), Saharanpur (224 sq.km.), Sitapur (201 sq. km.), Sultanpur (157 sq. km.) and Unnao District 197 sq. km. open forest area, while other districts are having less than 100 sq. km. open forest area which is insignificant. The open forest areas are found more along with the footsteps of Himalayas, along with the Chambal and Gomati rivers and Vindhyan range.

(iv) Total Afforestation

Total afforested area of Uttar Pradesh is 14,118 sq. km. which is 5.86 per cent to the total geographical area of the state and it is very insignificant than the 33 per cent earmarked forest cover for normal survival of the common people and restoration of ecological and environmental crisis of the state. Very high afforested area 2,469 sq. km. is seen in Sonbhadra District, followed by Kheri – 1,314 sq. km. while Shrawasti District has 811 sq. km. forest, Mirzapur – 782 sq. km., Pilibhit – 697 sq. km., Balrampur– 532 sq. km., Bijnor– 423 sq. km., Chandauli – 519 sq. km. and Chitrakut 554 sq. km. forest area while other Districts of Uttar Pradesh are having very nominal forest area.

It has been observed that there had been 4.46 per cent forest cover to the total geographical area of Uttar Pradesh during 1977. During 1999, the forest area percentage was 4.464 to the total geographical area having 0.002 per cent growth from 1977 to 1999. The percentage of forest area was found 5.705 during the year 2001, which has 1.24 per cent growth in the forest area from 1999 to 2001. It has been observed that there is 5.86 per cent forest area during 2003, to the total geographical area of the Uttar Pradesh. There is 1.398 per cent growth in the forest area during 1977-2003 in Uttar Pradesh.

Thus, the percentage area under forest to the total geographical area of Uttar Pradesh is very insignificant as per the National Forest Policy and Uttar Pradesh State Forest Policy, while percentage growth in the forest area is also very nominal which is a great concern to the common people and the state of Uttar Pradesh as well. The forest degradation and ecological and environmental crisis of Uttar Pradesh is a great challenge to the Government Departments/implementing agencies/ planners and the policy-makers of the state of Uttar Pradesh.

The most of our rural population is completely dependent on the forest resources to obtain fuel wood, fodder etc. There is explosive population growth in Uttar Pradesh along with the growth of cattle population on the one hand, and very insignificant growth in the forest area which is only 1.398 per cent from 1977 to 2003. The forest area of the Uttar Pradesh is only 5.86 per cent to the total geographical area in comparison to the 33 per cent forest cover earmarked by the National Forest Policy 1988 and Uttar Pradesh State Forest Policy 1998. The deforestation and land degradation crisis has to be controlled through people movement to meet the increasing demand of growing cattle and human population and also to restore the ecological and environmental crisis and also to improve the socio-economic crisis of the deprived rural multitudes.

4. Agro-afforestation in Sultanpur

Agro-afforestation programme in Sultanpur is not planned appropriately because it has no people's participation and need of the local people has not been given due importance for successful implementation, maintenance and management of the agro-afforestation activities in Sultanpur District. Various government departments/implementing agencies *viz.* Agriculture, Horticulture, Forestry departments of the District Sultanpur are engaged in the implementation of various programmes. Therefore, it is imperative to explain the department-wise agricultural-horticultural-afforestation programmes being implemented in Sultanpur district for sustainable development of the District Sultanpur.

Agricultural System

The agriculture is the main occupation of the people. Agriculture provides 74 per cent occupation to the people out of the total main workers in the District Sultanpur. The area under cultivation is 73.86 per cent followed by forest 0.44 per cent non-agricultural uses of the land 11.35 per cent. About 58 per cent of the agricultural land is irrigated through 2,044 km. long canals, 774 Government tube wells and other sources of irrigation. Wheat and rice are the major crops and pulses and sugarcane are the minor crops grown in the District Sultanpur.

Agricultural production plays an important role in the economic development of our country. Department of agriculture is contributing in the various agricultural and agricultural investment programmes. Presently, Rs. 6.68 lakh expenditure has been sanctioned under the District plan out of which Rs. 4.24 lakh expenditure has been allocated for high yielding variety of seeds, agricultural implements, irrigation pipes, insecticides and pesticides and also the field demonstrations through subsidy on agriculture to the farmers. Latest agricultural implements and agricultural techniques are being demonstrated to the interested farmers of the District Sultanpur.

In spite of the above 4691.20 mt. tonnes seed and 5432.25 mt. tonnes fertilizer has been distributed among the farmers and a sum of Rs. 2,873.35 lakh cropping loan and 59,340 farmers credit cards has been made available to the farmers. Thus, the department of agriculture is playing an important role for increasing the agricultural productivity and economic development of the District Sultanpur. Crop protection programmes play an important role in cropping system. Approximately, 25 to 30 per cent loss in the agricultural production is due to the insecticides starting from cultivation-crop cultivation, crop-harvesting and storage of the grain. The check and control of the loss in the cropping can be managed through insecticides and pesticides. The farmers are being technically trained and awareness training camps are being organised. Integrated Plant Protection Management demonstration and training programme is being demonstrated in the farmers' field. On an average 2,25,000 kg./litre insecticides, pesticides and chemicals and 4500 kg. bio-pesticides are being utilized for treating 5,25,000 hectares area of agricultural crops. Thus, it is observed that to increase the agricultural production in various scientific and technical support system and application of insecticides and pesticides are being made available to farmers of the District Sultanpur.

There are 65.8 per cent net sown area to the total area of the District Sultanpur. Proportionately very high percentage net sown area 70.5 per cent in Baldirai block and 70.1 per cent in Lambhua Block to the total area is recorded in the Baldirai Block of District Sultanpur. High percentage net sown area between 65 per cent to 70 per cent has been observed in the Jamo, Gauriganj, Amethi, Kurwar, Bhadainya, Dostpur, Akhandnagar, Pratappur Kamaicha, Kadipur and Motigarpur

Blocks of the District Sultanpur. Very low net sown area 55.1 per cent has been recorded in the Bhadar Block of the District Sultanpur.

Thus, it is observed that the agriculture is one of the main predominant occupations in the District Sultanpur because agricultural cropping system cover two-third area out of the total geographical area of the Sultanpur, but it requires agricultural scientific and technical support system to increase the agricultural production through multiple cropping system for optimum utilization of availavable land resources to meet the increating demand of growing cattle and human population in the form of fuel wood, wood, fodder, fibre, fruits, fisheries, vegetables and food grains etc. for sustainable development of the District Sultanpur.

Horticultural System

Horticultural cropping system of the District Sultanpur has been strengthened through newly horticultural plantation system. It has been planned to cover 1000 hectares area under horticultural plantation programme and one lakh fruit plants has been distributed among the horticultural growers of the District Sultanpur. Potato cultivation area has been covered in 5107 hectares and 305 quintals potato seeds has been distributed among the horticultural growers. Intensive potato cultivation techniques have been demonstrated in the 520 hectares land of the farmers.

Vegetable development programme has been implemented in the 26,500 hectares farmers land. The vegetables production target has been obtained 3,85,002 mt. tonnes. High yield variety seeds distribution obtained 52.42 quintals in the 1800 hectares hybrid seeds and horticultural techniques extension programme has been implemented in the horticultural growers field. Under the special component plan– fruits, vegetables – 4.5 hectares and new horticultural plantation – 20 hectares Scheduled Caste field has been practically demonstrated and special fruits development scheme has been launched in 110 hectares land of general cultivators and 20 hectares land of the Scheduled Caste farmers and the scheme has been launched in Amethi, Kurwar, Lambhua, Bhadainya development blocks of the District Sultanpur. Two horticultural nurseries have been established in Kadipur and Bhadar Blocks. In Kadipur Nursery – 6,640 grafted plants and 34,640 seedlings and in Bhadar nursery – 10,000 grafted plants and 32,000 seedlings nursery plants have been developed for the horticultural growers. In the Government Vegetables Field located at Dikhauali 5 hectares potato seed production has been ensured. Thus, it is recorded that horticultural cropping system has significant role in the sustainable development of the District Sultanpur.

There are 1. 67 per cent horticultural land to the total area of the District Sultanpur. Very high percentage of horticultural land 6.29 per cent has been recorded in the Sangrampur Block of the District Sultanpur. High percentage of horticultural land between 3.0 per cent to 5.0 per cent has been seen in the Jagdishpur, Amethi, Bhnetua, and Bhadar Blocks of the District Sultanpur. Medium percentage of horticultural land between 2 per cent to 3 per cent has been observed in Shukul Bazar, Baldirai, Shahgarh Blocks of the District Sultanpur while very low less than one per cent of horticultural land has been recorded in Dhanpatganj, Kurebhar, Jaisinghpur, Kurwar, Bhadainya, Dostpur, Akhandnagar, Lambhua, Pratappur Kamaicha, Kadipur and Motigarpur Blocks of the District Sultanpur. It is observed that area under horticultural programme in District Sultanpur is very insignificant. It requires integrated cropping system of Agri-Horti-Forestry programme of multicropping techniques for optimum utilization of available natural resources and human resources for sustainable development.

Afforestation System

The exploitation and destruction of forests in the District goes back to the British period where forest wealth were consumed for commercial gains. The trend for deforestation continued even after independence and the forests were used not for imperative economic growth but for other various reasons. The denudation of forest land, however, had now slowed down in the recent years despite human, cattle and commercial pressures due to efforts made by the various agencies. The growing cattle and human population, one of the main factors for deforestation, increases the need of the land for more food production to feed the growing number of people causes environmental degradation.

Afforested land in District Sultanpur is only 3.97 per cent to the total area of the District, which is very insignificant in comparison to the forest land earmarked by the National Forest Policy, 1988. Very high percentage of forest land area which is 4.62 per cent to the total area has been found in the Shukul Bazar followed by Musafirkhana 2.43 per cent forest land to the total area of the Block. Pratappur Kamaicha block has 1.33 per cent forest land to the total geographical area of the block. It has been observed that there are 0.66 per cent forest land in Kurebhar Block. 0.58 per cent forest land in Bhadainya Block, 0.29 per cent forest land in Jagdishpur Block, Dostpur 0.12 per cent and 0.17 per cent forest land in the Baldirai Block and Bhadar 0.44 per cent to the total area of the Block in the District Sultanpur. The Shahgarh, Gauriganj, Amethi, Dubepur, Lambhua and Motigarpur Blocks are having less than 0.09 per cent forest land to the total area of the Block, while Jamo, Bhnetua, Sangrampur, Kurwar, Akhandnagar and Kadipur Blocks are without forest land. The insignificant forest land in Sultanpur District is a great concern to the development workers, planners and administrators of the District Sultanpur. An integrated afforestation programme has to be launched in the district through people's participation and as per the local need. In view of the above and other facts, Government of India and other agencies have launched various promotional schemes all-over the country for afforestation and press tree plantation to increase the forest cover in which the participation of local people is also taking place. However, due to lack of centralized data base, the agencies are sometimes not in a position to plan their projects in an efficient way. The National Informatics Centre (NIC) Sultanpur had taken initiative to resolve their problem, though partially, by undertaking in the District Sultanpur. The report deals with the consolidation of the plantation work of various kinds of trees done by different agencies in different years geographically.

Social forestry division, Sultanpur, planted 2.53 lakhs plants in 118.50 hectares land, and 261.0 hectare area of the village community land has been planted, in which advance soil work, forest block and road side afforestation programme has been implemented and 1170 new Vikramgaurd has been constructed and trees have been planted.

There are 13 nurseries established by the social forestry division,

Sultanpur. Various varieties of useful plants have been developed in these nurseries. The developed plants of these nurseries are utilized for departmental plantation and larger afforestation programme. Out of the 13 nurseries, one nursery has been established in Amethi nearer to Railway Station. In the larger plantation programme, 52.69 lakh plants have been planted. The farmers of Sultanpur District have established 61 farmer's nurseries, in which approximately 60 lakh plants have been developed. The available plants from farmer's nurseries are being purchased for the plantation programme by the individuals and plantation agencies.

Bhumi Sanrakshan Adhikari Gomati (BSA) Sultanpur planted 70,875 plants covering 1055.50 hectares area in the District. The BSA Gomati planted 10,550 plants in 167.50 hectares land in Jagdishpur Block and 1500 plants has been planted in 214.0 hectares land in the Mursafirkhana Block while 29,100 plants has been planted covering 453.0 hectares land. BSA Gomati also planted 10,900 plants in the 154.0 hectares land in the Bhadainya Block while 5,325 plants has been planted covering 67.0 hectares land in the Kadipur Block of the District Sultanpur.

The District Forest Officer (DFO) Sultanpur, planted 9,03,430 plants covering 593.55 hectares land in Sultanpur. DFO Sultanpur has planted 86,400 plants in 53.25 hectares land is Shukul Bazar Block and 1,54,080 plants has been planted in 95.55 hectares land in Jagdishpur Block and 13,500 plants has been planted in the 10.0 hectare land in Musafirkhana Block of the Sultanpur District. There has been plantation of 17,400 plants covering 11.50 hectares land in Shahgarh Block and plantation of 36,350 plants in 28.50 hectares land in Gauriganj Block and 8,000 plants plantation in 5.0 hectares land in Amethi Block and 99,600 plants plantation covering 55.25 hectares land in Bhadar Block. There has been 16,000 plants plantation in 10 hectares land in Dhanpatganj Block and 78,400 plants plantation covering 43.25 hectares land in Jaisinghpur Block and 95,600 plants plantation in 53.25 hectares land in Kurwar Block and 1,32,200 plants plantation in 80.50 hectares land in the Dubepur Block and 20,000 plants plantation in 12.50 hectares land in Dostpur Block in the District Sultanpur. The District Forest Officer, Sultanpur planted 32,000 plants covering 70 hectares land in the Lambhua Block and 9,200 plants has been planted covering 5.0 hectares land in the Kadipur Block of the District Sultanpur.

The Indian Farms and Forestry Development Corporation (IFFDC) has planted 19,12,608 plants covering 1450.03 hectares land in Sultanpur District. It has planted 1,09,081 plants covering 79.55 hectares land in Shukul Bazar Block and 4,98,502 plants has been planted in 335.45 hectares land in the Jagdishpur Block while 2,99,287 plants has been planted in 248.27 hectares land in Baldirai Block and 1,81,465 number of plants has been planted in 139.75 hectares land in Shahgarh Block, while 1,33,922 plants has been planted covering 68.10 hectares land in the Bhadar Block. Thus, IFFDC Sultanpur has planted 75,512 plants covering 83.04 hectares land and 1,33,134 plants has been planted in 93.62 hectares land in Kurwar Block and 1,02,145 plants has been planted covering 74.20 hectares land in Dubepur Block and 81,933 plants has been planted covering 57.25 hectares land in Bhadainya Block while 2,97,627 plants has been planted covering 270.80 hectares land in Lambhua block of the District Sultanpur. Thus, Indian Farms and Forestry Development Corporation has planted 19,12,608 plants covering 10 Blocks of the District Sultanpur.

The District Horticulture Officer (DHO) Sultanpur has ensured plantation of 9,115 plants covering 100 hectares land is Sultanpur District. District Horticulture Officer, Sultanpur has ensured the plantation of 1,070 plants in 12.50 hectares land in Musafirkhana Block and 1,000 plants plantation has been done in 10 hectares land in Shahgarh Block and 1,200 plants plantation has been carried out covering 12 hectares land in Amethi Block and 850 plants plantation has been done in 10 hectares land in Dhanpatganj Block and 700 plants plantation has been carried out in 7 hectares land in Bhadainya Block of the District Sultanpur. The District Horticulture Officer, Sultanpur ensured the plantation of 1,350 plants covering 13.50 hectares land in Dostpur Block and 850 plants plantation has been doen in 8.50 hectares land in Akhandnagar Block and 630 plants has been planted in 9 hectares land in Lambhua Block and 525 plants has been planted in 7.50 hectares land in Pratappur Kamaicha Block while 940 number plants has been planted covering 10 hectares land in Kadipur Block of the District Sultanpur.

The Uttar Pradesh Land Development Corporation (UPLDC) has

carried out plantation of 59,509 plants covering 95.22 hectares land in Sultanpur District. It has planted 16,464 plants in 26.51 hectares land in Shahgarh Block and 31,450 plants has been planted in 40.52 hectares land in Bhnetua Block and 156 plants has been planted in 2.55 hectares land in Bhadar Block and 555 plants has been planted covering 0.89 hectares land in Dubepur Block and 5,340 plants has been planted in 8.55 hectares land in Bhadainya Block while 4,140 plants has been planted covering 7.20 hectares hand in Lambhua Block of the District Sultanpur. Thus, Uttar Pradesh Land Development Corporation has planted 59,509 plants in 6 Blocks of the District Sultanpur. The plantation programme in Sultanpur District has been very insignificant in comparison to the total geographical area of the Block.

Thus, *Bhumi Sanrakshan Adhikari,* Sultanpur, *District Forest Officer,* Sultanpur, *District Horticultrue Officer,* Sultanpur, *Indian Farms* and *Forestry Development Corporation* and *Uttar Pradesh Land Development Corporation* jointly planted 29.5 lakhs plants in the District Sultanpur. Very high 6.63 lakh plantation was done in Jagadishpur Block followed by 3.35 lakh Plantation in Lambhua Block followed by 2.58 lakh plants in Kurwar Block, 2.35 lakh plants in Bhadar Block, 1.95 lakh plants is Shukul Bazar Block and 2.16 plants in Shahgarh Block and 2.99 lakh plants in Baldirai Block and 2.34 lakh plants in the Dubepur Block, has been planted in Sultanpur District. Thus, there has been 1.05 lakh plants plantation in Sangrampur Block and 1.59 lakh plants plantation has been ensured in Jaisinghpur Block of the District Sultanpur. There has been 0.30 lakh plants plantation in Musafirkhana, 0.37 lakh plants in Gauriganj, 0.31 lakh plants in Bhnetua and 0.17 lakh plants in Dhanpatganj, 0.99 lakh plants in Bhadainya, 6.21 lakh plants in Dostpur and 0.15 lakh plants in Kadipur Block has been planted. Very insignificant plantation has been carried out in Pratappur Kamaicha Block (525 plants) Akhandnagar Block (850 plants) and Amethi Block (9,200 plants) of the District Sultanpur, while no plantation has been carried out in Jamo Block and Kurebhar Block of the District Sultanpur. It is observed that there has been done very least plantation in the various Blocks of the District Sultanpur which is very insignificant in comparison to the National Forest Policy 1988 and Uttar Pradesh State Forest Policy 1998.

Ecosystem of Sultanpur

(i) Flora

The major part of the district Sultanpur was covered with the forests of *Dhak* and thorny bushes forming a valuable site of refuge in trouble times of the Nawabi rules during former historical days. During the beginning of the 19th century, it is stated that one large tract of dense forest extended in an *Unbroken stretch* near *Ram Nagar* in *Tehsil Amethi.*

The Jungles of Bhadainya covered an area of more than 404 hectares even after the freedom struggle of 1857 when portions of Jungle tract were cleared off by the British forces. During the Second World War period and thereafter, in furtherance of the grow more food campaign, the jungles were recklessly cut down. Presently, small patches of jungles are seen near Unchgaon and to the west of Musafirkhana in the western part of the District Sultanpur. In the eastern part of the district, the remnants of extensive wood are found near the village of Kathaura, Rampur, Baragaon, Navgavan, Mahadeo and between Dulhapur and Bhavanpur. Small portions of land between Bhumi and Tikarmafi and the south of the Islampur, all in Tehsil *Amethi* are covered with natural vegetative growth. Small tract of jungle are also seen on the ravine lands of the banks of the Gomati River and Kadu Nala. The areas covered with timber and other trees and shrubs, which comes under the control of Forest Department is about 1600 hectares of which an area of about 1253 hectares lies in Tehsil Musafirkhana and 347 hectares in Tehsil Sadar Sultanpur. In addition to the above, road side avenues, controlled by the Forest Department are about 74 kms. in Tehsil Musafirkhana, 80 kms. in Tehsil Kadipur, 132 kms. in Tehsil Sadar Sultanpur and 60 kms. in Tehsil Amethi. The *jungle area* under the *control* of *Gram Sabha* is about *3,847 hectares* of which 1852 hectares are covered with timber trees and the remaining with other species of trees and shrubs. Of the timber jungles Tehsil Amethi contains 740 hectares, Tehsil Sadar Sultanpur contains 680 hectares, Tehsil Musafirkhana 394 hectares and Tehsil Kadipur 38 hectares. The forest area under other trees and shrubs cover 917 hectares in Tehsil Sadar Sultanpur, 527 hectares in Tehsil Musafirkhana, 325 hectares in Tehsil Amethi and 226 hectares in Tehsil Kadipur. However, these cannot be called forests of which they lack the stateliness and density, seen in the twilight at the season of the year their leaves are gathered for fuel; their crooked trunks and branches present the appearance of a number of gaunt, weired figures in all sorts of grotesque and fantastic shapes.

The main species of the trees found in these jungles are *Dhak, Shishan, Neem, Babool, Bel, Pipal, Bargad, Goolar*, *Pakar*, and *Mahua.* Among the species which have been introduced recently *Mango,* Khair, *Safed siris, Kala siris, Kachnar, Amaltas, Jamun, Sagaun, Semal, Arjun, Bahera* and *Zezyphus* are commonly seen in the area under the forest department along the roadsides in the District Sultanpur.

(ii) Fish

Fish are found in the rivers, lakes, ponds, canals and artificial reservoirs of the District Sultanpur. The main species of the fish found are *bata, rehu, karuanch, singhi, nain, raia, bhakur* and *belgagra.*

As per the policy of the Government, low-lying or the wastelands or common village Panchayat land is being allotted to the interested needy people on *Patta* for fisheries. To improve the village pond for fisheries, a sum of Rs. 60,000 per hectare and a sum of Rs. 30,000 per hectare is being made available on loan to the fisherman for the initial investments of the fishing. Out of the sanctioned loan 20 per cent grants for general person and 25 per cent subsidy for the Scheduled Caste person is being provided to the fisherman.

For the construction of pond on the individual land, Rs. 2 lakh per hectare and for the initial inuestment of fishing Rs. 30,000 per hectare bank loan is being made available, while 20 per cent for the general and 25 per cent for the Scheduled Caste person subsidy is being provided to the above loan.

The animal husbandry programme *viz.* cow, buffalo, duck, hen and pig along with fisheries can be managed and along with the bunds of the pond fruits and vegetables cultivation can be implemented to increase the income of the fishermen and to ensure optimum utilization of the pond land. For the integrated fisheries programme, a sum of Rs. 80,000 per hectare is also being provided by the bank out of which 20 per cent subsidy for general person and 25 per cent subsidy for the Scheduled Caste person is being made available by the Fisheries Development Agency, Sultanpur.

As per the Government order, individual landowner who has constructed pond in their own land and engaged in the fisheries programme, general person 20 per cent and Scheduled Caste person 25 per cent subsidy is being made available to the total cost estimated for the construction of pond and fisheries initial investment during the first year of the fishing work.

Scientific training-cum-demonstration programme for fish cultivation for 10 days short duration is being provided by the Fisheries Development Agency and during such training period Rs. 100 as training expense and Rs. 100 as tour expense at once is also being provided to the fishermen.

For the individual fish hatchery programme up to the 1.5 hectare pond for mini-hatchery construction for 10 million fry capacity fish hatchery, a sum of Rs. 8 lakh loan is being provided and at the rate of 10 per cent subsidy maximum up to Rs. 80,000 is also provided to the fishermen.

As per the demand of the fishermen, rohu, bhakur, nain, kamal carp, sieve carp and grass carp high-tech fish seedlings are being made available to the hatcheries fishermen from the fisheries department's pond by the Uttar Pradesh Fish Development Corporation at the nominal government's rates.

The soil and water testing is being ensured free of cost for the ponds of the fishermen by the Fisheries Development Agency.

The scientific technological advice is being made available time to time to the fishermen as per growth of the fish.

For the integrated development of fishermen Fishermen Cooperative Societies are formed for the self-sustenance of the fishermen.

Under the scheme of Fishermen Accident Insurance, the registered Fishermen Cooperative Societies are able to get the insured amount ratio by the Government of India and the Government of Uttar Pradesh, in this programme Fishermen Co-operative Societies members during the accidental conditions get Rs. 50,000 in case of death and Rs. 25,000 in case of handicapness insurance amount is being made available by the Fisheries Development Agency.

In the fishermen dense populated areas, the below poverty line fishermen are being provided Rs. 25,000 for housing and for 10 houses jointly Rs. 20,000 is being provided for installation of hand pump under the scheme of the Government.

(iii) Fauna

The wildlife of the district has greatly decreased in number and variety since the middle of the 19th century due to the clearance of jungles and the reclamation of wild tracts and groves for cultivation. Though the wild animals have become quite unimportant, yet considerable species of birds, reptiles and fish are found in the District Sultanpur.

(iv) Animals

The stray leopard which was occasionally seen in the jungles of Kadu Nala has now become stinct. The wolf has become scarce. The Nilgai is seen is the jungles near Ramnagar and few other places. Besides the monkey, jackal, fox and hare are common animals found throughout the District Sultanpur.

(v) Birds

The birds of the District Sultanpur are similar to those of the adjoining districts. The chief game birds found in the district are of several varieties of wild goose, duck, quail and partridge which are fairly plentiful during the winter. The large egret which is found throughout the district is shot for the sake of its plumage. Among other birds generally found in the District Sultanpur are parrot, peacock, red jungle fowl, nightingale and sparrow. A large number of migratory birds come to the swamps and jhils in the southern part of the District Sultanpur.

(vi) Reptiles

Various varieties of snakes and other reptiles are found everywhere in the District Sultanpur in the rural areas, Harmless snakes are found but some are deadly, e.g. the cobra. The russel's viper which is viviparous and nocturnal in its habits is commonly found in the District Sultanpur. Though the majority of snakes are non-poisonous, a few people die of snake bite almost every year. The other reptiles found in the District Sultanpur are chameleon and monitor lizard, the later which was the fast becoming extinct due to netting, shooting, has been declared protected species in the Sultanpur District.

Thus, the ecosystem of the area has been degraded which has led to ecological environmental crisis and socio-economic backwardness.

5. Conclusion

The land degradation and deforestation has caused severe ecological, environmental crisis and socio-economic backwardness. The symbiotic relationship between man and nature has been realized because the poor depends on the forest. The Scheduled Caste communities totally depend on the minor forest produce, to grass and fallen dry wood for fuel wood. The relationship between man and nature has been threatened. Agro-afforestation efforts, an endeavour is rendered more difficult since the needs of the community are at a variance with each other and it cannot be reconciled unless poverty amelioration programmes raise the level of living of those who are surviving below poverty line. The agro-afforestation programme must be initiated as a people's movement.

Agro-afforestation is a system of land use along with the combined growing of agricultural crops with social forestry, horticulture, animal husbandry including vegetables, fuel wood, fodder, fibre, fruits, fisheries and food grains etc. to fulfil the increasing demand of growing cattle and human population.

There is tremendous pressure of the human population which has led to deforestation and land degradation. The stabilization of population and participatory management for the reclamation of wastelands and agro-afforestation management for sustainable development.

The Forest Survey of India, Dehradun indicated in its report that there are 5.86 per cent forest land in Uttar Pradesh. Various programmes have been launched by the Government to ensure successful afforestation programmes. There are 1297 sq. km highly dense forest land out of which 210 sq. km highly dense forest is found in Shrawasti District followed by 144 sq. km in Balrampur District. There are 4,699 sq. km dense forest in Uttar Pradesh out of which 846 sq. km dense forest is recorded in Sonbhadra District followed by 502 sq. km in Kheri, 346 sq. km in Chitrakoot and 316 sq. km dense forest is found in Mirzapur

District of Uttar Pradesh. There are 8,122 sq. km open forest in Uttar Pradesh, out of which 1606 sq. km. open forest is found in Sonbhadra District followed by Kheri – 446 sq. km, Shrawasti – 347 sq. km, Chandauli – 327 sq. km open forest area. The forest land is found more along with the foot steps of Himalayas and Vindhayan range and the Chambal and Gomati rivers.

Thus, in Uttar Pradesh, there are 14,118 sq. km forest area which is 5.86 per cent to the total geographical area of the state which is very insignificant in comparison to the 33 per cent forest land earmarked by the Government. Very high afforested area is found in Sonbhadra District – 2,469 sq. km followed by Kheri -1,314 sq. km forest land and Shrawasti District – 811 sq. km, Mirzapur – 782 sq. km, Pilibhit– 697 sq. km, Balrampur 532 sq. km, Bijnor – 423 sq. km, Chandauli – 519 sq. km and Chiarkoot 554 sq. km forest area. It has been recorded that during 1977, there was 4.46 per cent forest land and during 2003 there has been 5.86 per cent forest area in Uttar Pradesh having 1.398 per cent growth in the forest land from 1977 to 2003, which is insignificant against 33 per cent forest land earmarked by the Nation Forest Policy and Uttar Pradesh State Forest Policy. The land degradation and deforestation is a great challenge to the planners and programme implementing institutions.

The agro-afforestation programme in Sultanpur is not appropriately planned and properly implemented by the implementing agencies, because it requires joint endeavour and efforts of Agriculture, Horticulture, Social Forestry and Fisheries Departments of the District Sultanpur with the involvement of farmers.

The agriculture is the main occupation of the people of Amethi and Sultanpur. The area under cultivation is 73.86 per cent in Sultanpur followed by 0.44 per cent forest area.

The agriculture department has expended Rs. 6.68 lakh for high yielding variety seeds, fertilizers, agricultural implements, irrigation pipes, insecticides and pesticides and field demonstrations. Latest developed agricultural implements, agricultural cropping etc. are practically demonstrated in the farmers' field. In spite of the above 4691.20 mt. tonnes seed and 5432.23 mt. tonnes fertilizer has been distributed among the farmers and a rum of Rs. 2873.35 lakhs cropping loan and 59,340 farmers credit cards has been also provided farmers. It has been observed that about 25 per cent to 30 per cent loss in the agricultural production is due to the insecticides starting from cultivation — cropping — crop harvesting and storage of the grain. Appropriate application of insecticides and pesticides can reduce the loss in agricultural production.

Sultanpur District has 65.8 per cent net sown area. Proportionately high percentage more than 70 per cent net sown area is found in Baldirai Block and Lambhua Block of the District Sultanpur while very low percentage 55.1 per cent net sown area to the total area is found in Bhadar Block of the District Sultanpur.

Horticultural cropping system has been strengthened in Sultanpur district covering 1000 hectares land under horticultural plantation. Intensive potato cultivation has been ensured in 5107 hectares land and 305 quintals potato seed has been distributed and high-tech potato cultivation has been demonstrated in 520 hectares farmers land. Vegetable development programmes are being launched in 26,500 hectares farmers' land providing 52.42 quintals seeds distributed and in the 1800 hectares land horticultural techniques has been demonstrated in the farmers' field. Horticultural nurseries has been established through developing 6,640 grafted nurseries plants at Kadipur and 34,640 grafted plants seedlings and about 10,000 grafted plants and 32,000 seedlings nurseries plants has been developed at Bhadar Nursery. There are 1.67 per cent horticultural land in Sultanpur District. Very high percentage 6.29 per cent horticultural land is found in Sangrampur Block and 3.41 per cent horticultural area is seen in Amethi Block of the District Sultanpur.

The afforested land in District Sultanpur is 3.17 per cent to the total area. Very high percentage 4.62 per cent forest land is found in Shukul Bazar Block followed by Musafirkhana having 2.43 per cent forest land and Pratappur Kamaicha 1.33 per cent forest land to the total area of the Block while Amethi Block is having only 0.05 per cent forest area to the total area of the Block which is very insignificant.

Social Forestry Division Sultanpur planted 2.53 lakh plants coverning 118.50 hectares land and 261.0 hectares land belonging to Gram Panchayat. There are 13 nurseries established in Sultanpur while one nursery at Amethi has also been established. In Amethi Block, 8000 plants has been planted covering 5 hectares land and 9.03 lakh plants has been planted covering 593.55 hectares land during 1996-98 in Sultanpur District.

Bhoomi Sanrakshan Adhikari (BSA) planted 70,875 plants covering 1055.50 hectares land.

District Horticulture Officer planted 9,115 plants covering 100 hectare land in Sultanpur while 1200 plants has been planted in 12 hectares land. Indian Farm and Forestry Development Corporation (IFFDC) planted 19.13 lakh plants covering 1450.03 hectares land in Sultanpur District while Uttar Pradesh Land Development Corporation planted 59,509 plants covering 95.22 hectares land. Thus, in Sultanpur District 29.56 lakh plants has been planted and 9,200 plants has been planted covering 17 hectares land in Amethi Block.

The major part of the Sultanpur District was covered with the *Dhak*, thorny bushes. A large dense forest tract extended in an unbroken stretch near Ram Nagar in Tehsil Amethi. The jungles area under Gram Sabha is about 3,847 hectares and 1,600 hectares forest land is controlled by the Forest Department, while 325 hectares forest land is found in the Tehsil Amethi.

The main species of trees found in the jungles are *Dhak, Shisham, Neem, Babool, Bel, Pipal, Bargad, Goolar, Pakar* and *Mahua*. Among the species which has been introduced recently are *Mango, Khair, Safed Siris, Kala Siris, Kachnar, Amaltas, Jamun, Sagaum, Semal, Arjun, Bahera, Zezyphus* are commonly seen in the area under the forest deparments along the reoadsides in the Sultanpur District.

The fish are found in the rivers, lakes, ponds, canals and artificial reservoirs of the District Sultanpur. The main species of the fish found in the district are *bata, rehu, karuanch, singhi, nain, raia, bhakur* and *belgagra*. The fish culture is scientifically and technically promoted and demonstrated in Sultanpur District under various schemes launched by the Fisheries Development Agency, Sultanpur.

The flora and fauna and wildlife in Sultanpur District has greatly decreased in number, which are unimportant in the area.

The ecosystem of the area has been degraded which has led to ecological, environmental crisis and socio-economic backwardness of the area.

Thus, agro-afforestation programmes in Sultanpur District and in Amethi Block has not been properly planned and appropriately implemented because it requires joint venture and efforts of the implementing agencies like Agriculture – Horticulture – Social forestry and fisheries and with the participation and involvement of the local people. The agro-afforestation programmes should be implemented for the people, by the people and demonstrated in the farmers' field in an integrated manner adopting latest developed and innovated scientific techniques.

3

Factors in Wastelands

Hridai R. Yadav

1. Introduction

The spatial distribution of wastelands and the marked changes in the areas of wastelands in Amethi and Sultanpur District necessitates to probe into the cause and effect phenomenon of wastelands. Various causative factors responsible for the development of wastelands are numerous but the predominant ones can be identified and understood easily.

The development and formation of wastelands as a type of land use system can be attributed to a complex processes interacting at varying levels. Therefore, it has become necessary to investigate the factors affecting wastelands of Amethi Block at village level and also at the Block level in Sultanpur in general. It is assumed that development of wastelands occurs through unhealthy interaction of triangle of agencies *viz.* Man → Nature → Technology. In other words, where the environment is severe, man's ability to interact with nature to his advantage is curtailed and man fails to fully avail himself of the land resources. On the other hand, the combined impact of interaction between man and nature, when technology as a tool is very low, wastelands development grows on the whole, the combined impact of interaction between man and nature and varying levels of technological inputs are considered to be the main factors involved in the formation of wastelands. Though, the natural factors assume primary importance, man's varying levels of ability to utilize land through technological innovations also accelerate wasteland formation. Therefore, it is necessafy to clearly identify and investigate all determinants of wasteland formation.

In view of the above assumptions, various natural (morphometric, hydrologic, pedologic) and human anthropological factors have been selected for analyzing each wasteland type separately, because generation and changes cannot be made for all the wasteland types taken together for the reason that different factors have varying impact on them. The major factors have been grouped into two main heads and sub-divided as follows :—

I. Natural Factors

(a) Morphometric

1. Slope
2. Ruggedness number

(b) Hydrologic

3. Drainage Density
4. Quality of water in pH
5. Behaviour of water-table

(c) Pedogenic

6. NPK of the Soil
7. pH value of the Soil

II. Human Factors

8. Land concentration (Gini's co-efficient Ratio)
9. Percentage of Scheduled Caste Population
10. Population Growth
11. Growth of Fertilizer
12. Growth in Gross Irrigated area
13. Agricultural workers.

Thus, the causative factors responsible for the development of wastelands are numerous but the major and predominant natural and human factors can be identified and analysed with the view to check and control such factor for further strategies. Keeping the above facts into consideration, an attempt has been made to analyse various human factors, to analyse its impact on wastelands development, in general so that the block level development of wastelands can be analysed for the formulation of planning strategies to reclaim the wastelands for agro-afforestation.

2. Factors of Wastelands in Sultanpur

The relationship between man and nature has been shattered. The nature has given bounty to man but in return man could not do anything for restoration and sustenance to nature. Man exploited the nature excessively due to which ecological and environmental crisis has taken place, which has caused socio-economic crisis. The combined impact of interaction between human and nature and varying levels of technological inputs are considered to be the predominant factors responsible for the development of wastelands at different levels. Out of numerous causative factors, some of the important anthropogenic or human factors are explained at Block level variation of indicators in Sultanpur district.

Population Growth

The explosive population growth is one of the predominant factors in the development of wastelands. The natural land resources are limited and man or population is totally dependent of natural/land resources. The overexploitation of natural land resources to meet the increasing demand of fuel wood, fodder, fibre, fruits, fisheries and food grains etc. has caused the land degradation causing ecological and environmental crisis.

The population growth in Sultanpur District has been recorded 25.27 per cent from 1991 to 2001. The population growth during 1991 to 2001 has been observed very high in Jagdishpur Block having 47.03 per cent. This population growth has been recorded very high due to industrialization in Jagdishpur area and people's immigration. The excess/explosive population growth and industrialization has caused development of wastelands in the study area. High population growth between 25 per cent to 30 per cent has been found in Shukul Bazar, Amethi, Kurebhar, Dubepur, Bhadainya, Lambhua, Pratappur Kamaicha, Kadipur Block of Sultanpur District. The population growth in these blocks have positive growth due to urbanization and small-scale secondary and tertiary workers, concentration which has caused positive change in the development of wastelands. Very low percentage population growth less than 20 per cent has been recorded in Bhnetua, Sangrampur and Dhanpatganj Blocks.

Thus, it has been noticed that the excessive population growth play an important role in the development of wastelands because the infrastructional, institutional and peoples' need based development programmes are one of the major factors in the development of wastelands.

Scheduled Caste Population

The proportionate distribution of Scheduled Caste population also plays an important role in the development of wastelands. The Scheduled Caste population has positive impact on wastelands development because the Scheduled Caste population is mainly dependent on natural land resources and the overexploitation of natural land resources causes land degradation and wastelands development. The Scheduled Caste population in Sultanpur District is observed 22.9 per cent to the total population of the district. It has been observed that very high percentage of Scheduled Caste population to total population is found in Jamo Block having 30.1 per cent Scheduled Caste population. High percentage of Scheduled Caste population between 25 per cent to 30 per cent is found in Shukul Bazar, Musafirkhana, Gauriganj, Akhandnagar, Kadipur and Motigarpur Blocks of the Sultanpur District. Very low percentage of Scheduled Caste population less than 20 per cent is recorded in Amethi, Sangrampur, Kurwar and Dubepur Blocks of Sultanpur District.

In view of the above, it has been observed that the Scheduled Caste population has positive impact on wastelands formation, because the Scheduled Caste population in the remote rural areas are mainly dependent on natural land resources in the form of fuel wood, fodder, fibre, fruits, vegetables, fisheries, wood, food grains etc. The over-exploitation of the land resources in an unorganized manner to fulfil the increasing demand of the growing people has led to land degradation and wastelands development.

Fertilizer Growth

The application of fertilizer plays an important role in the formation of wastelands. The cultivators have common feeling that high doses of fertilizer application will increase the agricultural production of the field. In anticipation of increasing agricultural productivity, the cultivators are applying fertilizer in a multiple system, which at a certain stage in spite of increasing agricultural productivity, reduces the fertility status of the soil. Thus, expenditure in agricultural production is higher than the profit accrued out of agricultural farming. In view of the low profit than high expenditure farmers have left the land uncultivated, and such land is lying as wasteland. The higher doses of fertilizer application or growth in fertilizer use has positive impact on wastelands development. The percentage growth of fertilizer application has been computed from 1991 to 2001. The fertilizer growth in Sultanpur District has been noticed as 37.8 per cent. Very high percentage fertilizer growth, i.e. 47.8 per cent has been observed in Kadipur Block while high percentage fertilizer growth between 40 per cent to 45 per cent has been recorded in Shahgarh, Sangrampur, Dostpur Blocks of District Sultanpur. Very low percentage fertilizer growth less than 15 per cent is found in Dhanpatganj, Dubepur, and Bhadainya Blocks while low percentage fertilizer growth between 15 per cent to 20 per cent is noticed in Akhandnagar Block of Sultanpur District. Thus, it is observed that fertilizer application growth is very significantly distributed at Block level in Sultanpur District. It is noticed that the high doses of fertilizer application has positive impact on wastelands development. Therefore, it is necessary that before application of fertilizer the soil status should be tested and as per the soil requirements the organic and inorganic amendments should be applied for various crops in different cropping pattern.

Growth of Gross Irrigated Area

High doses of irrigation water is injurious to the soil because it disturbs the water-table and fertility status of the soil percolates with underground water and it creates problem of waterlogging. Thus, it is assumed that high doses of irrigation water and percentage growth in gross irrigated area has positive impact with wastelands development. There is 15.2 per cent growth in Gross Irrigated Area from 1991 to 2001. Very high percentage of growth in gross irrigated area is found in Musafirkhana Block which is 36.6 per cent from 1991 to 2001. High percentage growth in gross irrigated area between 20 per cent to 30 per cent is observed in Jagdishpur, Shahgarh, and Gauriganj Blocks of District Sultanpur. Very low percentage of growth in gross irrigated area less than 10 per cent is recorded in the Sangrampur Block while low percentage growth in gross irrigated area is noticed in Baldirai, Jamo, Amethi, Bhnetua, Bhadar, Kurwar, Bhadainya, Dostpur, Akhandnagar, Pratappur Kamaicha and Motigarpur Blocks of District Sultanpur. Thus, high doses of irrigation and growth in gross irrigated area has positive relationship with wastelands development in District Sultanpur.

Agricultural Workers

Proportionate distribution of agricultural workers have positive impact with the wastelands development. Very high percentage of agricultural workers more than 85 per cent is found in Akhandnagar Block while high percentage between 80 per cent to 85 per cent agricultural workers are recorded in the Jamo, Shahgarh and Dostpur Blocks of District Sultanpur. Very low percentage of agricultural workers less than 60 per cent is found in the Dubepur Block while low percentage agricultural workers between 60 per cent to 70 per cent are observed in Jagdishpur, Kurebhar, Kurwar and Dubepur Blocks of District Sultanpur. Thus, there is positive relationship between agricultural workers and wastelands development at Block level in Sultanpur District.

Population in Sultanpur

The total population of District Sultanpur is 31,30,926 persons comprising 16,11,936 males and 15,78,990 females. The 4.77 per cent urban population while rest of them are the rural population. Agricultural workers, household industry workers, other workers and marginal workers are very insignificant while very high percentage of non-workers are observed in Sultanpur District. In the urban areas other workers and marginal workers are significantly found while more than 75 per cent non-workers are found in rural and urban and rur-ban areas as well. The population distribution in Sultanpur District indicates that there is positive relationship with the wastelands development in Sultanpur District.

Industries

It has been observed that industries have positive impact on wastelands. The agricultural, metallurgical (iron and steel), heavy engineering, light engineering, cement, sugar, handicraft, handloom industries, paper and paper board industries are major industries. The Sultanpur District has many favourable indicators for industrial development. There are 11 major and medium industries like paper, cement, sugar, plastic, fertilizer, iron and aeronautical have been established mainly in Jagdishpur, Tikaria, Munshiganj, industrial areas of the District Sultanpur. Brassware, textile, agricultural implements, wood and steel, furniture, oil and floor, soap, footwear, stings and ropes, pottery, jaggery, and cardboard boxes are the small and household industries existing in different areas of District Sultanpur. It has been observed that there is positive relationship between industrial development and wastelands development in District Sultanpur.

3. Factors of Wastelands in Amethi

The factors for wasteland development at village level has been analysed through natural and human indicators. The natural factors at grass roots level really play an important role in the formation of wastelands.

Various morphometric, hydrologic and pedogenic factors and anthropogenic indicators have been analysed to explain the relationship between wastelands formation and natural and human factors. Attempt has been made in this study to analyse the relationship between dependent and independent variables adopting statistical methods, i.e. correlation matrix and step-wise regression analysis. The accumulated information has been digitized and mapped through Geographical Information System (GIS) at village level for Amethi Block.

Slope

Very high slope more than 1.50^0 has been observed in Naraini, Mochawa, Kohra, Bhusahari and Gaderi villages of Amethi Block, while high slope between 1.00^0 to 1.50^0 is recorded in Kakawa, Hathkila and Ramgarh villages of the Amethi Block. Very low slope less than 0.60^0 has been found in Katarafulkunwar, Mahso, Rebha, Mahmudpur, Saraikhema, Katara Maharani, Trilokpur, Tala, Kushi Tali, Korari Girdharshah, Nuanwa, Umapur Ganapatti, Dhandhudhar, Mochwa, Agahar, Himmatgarh, Nainaha Bartali and Gaderi villages of the Amethi Block of Sultanpur District. It has been observed that slope has positive relationship with the wastelands development, because higher degree of slope causes more soil erosion which leads to the deterioration of the fertility status of the soil and at a later stage the cultivators oftenly use leave the land uncultivated and such land is left as wastelands. The soil erosion and deterioration of fertility status of the soil is responsible for the ecological and environmental degradation.

Ruggedness Number

Very high ruggedness number more than 0.38 is found in Naraini and Ramdaipur village, while high ruggedness number between 0.25 to 0.30 has been seen in Benipur, Hathkila, Raidaipur, Loniapur, Kokra, Maharajpur and Ramgarh villages of Amethi Block. Very low ruggedness number has been found in Rebha, Tala Himmatgarh, Nainha Bartali and Gangauli villages of Amethi Block. Low ruggedness number has been observed in Kherauna, Parsanwa, Katarafulkunwar, Dedhpasar,

Mahso, Raipurfulwari, Mahmudpur, Saraikhema, Katara Maharani, Trilokpur, Kushitali, Korarigirdharshah, Nuanwa, Umapur Ganapatti, Dhandhudhar, Mochwa, Mahmudpur, Agahar and Dehra villages of Amethi Block. It has been observed that there is positive relationship between ruggedness number and wastelands development. Proportionately high ruggedness number will lead towards high wastelands development.

Drainage Density

Very high drainage density more than 4.50 par sq km has been seen in the Loniapur, Dedhpasar, Saraikhema, Saraiya, Duban and Maharajpur villages of Amethi Block. High drainage density per sq km from 4.00 to 4.50 has been recorded in Parsanwa, Benipur, Hathkila, Dedhpasar, Rebha, Loniapur, Raipurfulwari, Mahmudpur, Jangal Ramnagar, Darkha, Dhandhudhar, Chaturbhujpur, Mochwa, Kakwa, Dehra, Nainha Bartali, Bhusahari and Gangauli villages of the Amethi Block. Very low drainage density below 3.00 per sq km has been found in Mahmudpur and Himmatgarh villages of the Amethi Block. It has been observed that there is positive relationship between drainage density with wastelands development. The high number of drainage system causes soil erosion and water accumulation which develops the wastelands formation in the area.

Quality of Water in pH

There is positive relationship between quality of water in pH and wastelands development. High pH value in the quality of water leads towards salinity and it develops usarisation in the soil hence develops the area in wastelands. Very high quality of water in pH more than 9.0 has been seen in the Hathkila village and high quality of water in pH between 8.5 to 9.0 has been recorded in Dedhpasar, Loniapur, Mahmudpur, Saraikhema, Jangal Ramnagar and Dhandhudhar, Saraiya Duban villages of the Amethi Block. Very low quality of water in pH less than 7.75 has been observed in the Raidaipur, Ramdaipur, Bhaganpur, Korarigirdharshah, Nuanwa, Mochwa, Kohra, Mahmudpur,

Purabgaon, Himmatgarh, Dehra and Gaderi villages of the Amethi Block.

Behaviour of Water-Table

The high water-table in fact between 22 to 24 feet has been seen in the Hathkila, Dhandhudhar, Saraiya Duban, Maharajpur and Nainaha Bartali villages while very high behaviour of water-table more than 24 feet has been observed in the Dedhpasar, Loniapur, Saraikhema villages of the Amethi Block. Very low behaviour of water-table less than 14 feet has been found in the Raidaipur, Tala, Bhaganpur, Nunawa, Himmatgarh and Dehra villages of the Amethi Block. It has been assumed that there is positive relationship between behaviour of water-table and wastelands development.

NPK of the Soil

It has been assumed that there is negative relationship between NPK of the soil and wastelands development, because the low NPK percentage of the soil will cause the low agricultural productivity and increase the expenditure/cost of the production due to which the cultivators are forced to lease the land uncultivated. Very high NPK of the soil more than 3 per cent has been observed in the Kherauna, Benipur, Mahmudpur, Saraikhema and Nuanwa villages while high (2.5 to 3.0) NPK of the soil has been found in the Dedhpasar, Mahso, Raidaipur, Ramdaipur, Jangal Ramnagar, Katara Maharani Kushitali, Loharata, Korarigirdharshah, Loharta, Kohara, Saidpur, Mahmudpur, Himmatgarh, Dehra, Bhusahari villages of the Amethi Block. Very low NPK of the soil less than 2.10 per cent has been observed in the Agahar, Maharajpur villages of the Amethi Block of Sultanpur District.

pH Value of Soil

There is positive relationship between pH value of the soil and wastelands development. The very high pH value of the soil will lead to usarisation of the soil which leads towards much formation of wastelands development. Very high pH value of the soil more than 8.5 has been recorded in the Hathkila, Loniapur, Saraikhema, Umapurganapatti, and Saraiya Duban villages of the Amethi Block. Very low pH value of the soil less than 7.6 has been found in the Raidaipur, Tala, Kushitali villages of the Amethi Block.

Land Concentration

The size of landholding has been computed adopting Gini's co-efficient ratio formula to find out the concentration of size of landholding. The land concentration has positive relationship with wastelands development because high size landholders and land concentrated cultivators do not care much attention for low productive land due to which the area under wastelands has been increased. Very high land concentration more than 0.8 has been found in the Jangal Ramnagar, Saidpur, Agahar, Himmatgarh and Bhusahari villages of the Amethi Block. Very low land concentration is less than 0.5 has been recorded in the Benipur, Rebha, Ramdaipur, Nunawa and Dehra villages of the Amethi Block of the District Sultanpur.

Scheduled Caste Population

The Scheduled Caste population has positive relationship with the wasteland development, the Scheduled Caste population are oftenly the landless population. If the percentage of Scheduled Caste population in area is high then the land will be concentrated among the limited households. The household having concentration of land oftenly do not pay attention for the less productive land or less fertile land which causes an increase in the area under wastelands. Very high percentage of Scheduled Caste population more than 35 per cent has been observed in the Kushitali, Ramdaipur villages while high percentage of Scheduled Caste population between 30 per cent to 35 per cent has been found in Saidpur and Gaderi villages of the Amethi Block. Very low percentage of Scheduled Caste population less than 10 per cent has been recorded in the Dedhpasar, Bhaganpur and Dehra villages of the Amethi. It has been observed that the Scheduled Caste population in Amethi Block at village level has been significantly distributed.

Thus, it has been observed that the NPK of the soil has negative relationship with the wastelands development, while slope in degree, ruggedness number, drainage density per sq km quality of water in pH, behaviour of water-table in feet, pH value of the soil, land concentration and Scheduled Caste population have positive relationship with wastelands development in the selected area of the study.

4. Analysis of the relationship between natural and human factor with Wastelands Development

The relationship among the dependent and independent variables has been analysed through the *Correlation Matrix* and *Step-wise Regression* analysis.

The correlation matrix has been computed to see the trend of the correlation among the different independent variables for various types of wastelands development. Step-wise regression analysis has been computed to see as how the parameters change when new variables are added one by one.

Waterlogged Land

The correlation matrix indicates that there is significant positive correlation between drainage density (x_3) independent variable and proportion of waterlogged land (y) dependent variable which is significant at 1 per cent level of significance, because high drainage density helps in accumulation of surface water in the low-lying areas, which is helpful in the rise of water-table of the area.

The positive correlation between slope (x_1) and ruggedness number (x_2) which is significant at 5 per cent and 10 per cent levels of significance respectively. High values of ruggedness number and slope creates positive conditions for maximum surface waterflow and accumulation of water in the low-lying areas. Behaviour of water-table (x_5) has positive correlation with the waterlogged land but this relationship is significant at 5 per cent level of significance. There is negative correlation between NPK of the soil which is significant at 10 per cent level of significance because low NPK of the soil means low fertility status of the soil. The other variables *viz.* pH value of the soil (x_8), land concentration (x_7), Scheduled Caste population (x_9) and quality of water represented in pH value (x_4) has positive correlations with the waterlogged land but these relationships are insignificant. Therefore, these variables fail to provide explanation for variation in the proportion of waterlogged land at village level in the Amethi Block of the Sultanpur District.

The step-wise regression analysis has been computed to study the correlation of various dependent variables with the proportion of water-logged land as the independent variable. The most predominant variable in explaining the variability in waterlogged land is found to be the slope (x_1) which explains 29.9 per cent variation in the waterlogged land at village level in the Amethi Block. It is also interesting to note that in the subsequent step, the variable drainage density (x_3) enters and jointly with variable x_1 explains 32.7 per cent total variation in the proportionate distribution of waterlogged land.

In the step-wise regression analysis, the other variables could not enter in this system of analysis. It indicates that these variables fail to explain the variation in the proportionate distribution of waterlogged land at village level in the Amethi Block of the District Sultanpur.

Usar Land

It has been observed that there is a positive correlation between usar land and pH value of the soil (x_7) which is significant at 1 per cent level of significance. It indicates that the villages having proportionately high percentage of usar land have high pH value of the soil because of excessive concentration of neutral salts mainly chlorides and sulphates of sodium in usar soils. It has been observed that there is positive correlation between land concentration (x_8) and Scheduled Caste population (x_9) with the distribution of usar land at village level in Amethi Block which are significant at 5 per cent level of significance because the farmers having large size of landholdings are unable to remaintain the fertility status of the soil, while such cultivator pay much attention for cropping in the fertile land because the farmers are quite certain to achieve due benefit from such cropping system and the Scheduled Caste farmers in most of the areas are landless but some of the Scheduled Caste people have been allotted wastelands on *patta* basis but such usar land is not yet reclaimed by the Scheduled Caste farmers due to ignorance and socio-economic backwardness due to which the usar lands are still left out of cultivation. The behaviour of water-table (x_5) has positive correlation with the user land which is significant at 10 per cent level of significance. In the usar land water-table is low and at a certain strata the soil profile is not very preamble due to the presence of *kankar pan*. Due to this the water-table rises and water gets enriched through soluble salts. It has been recorded that there is negative correlation between the NPK of the Soil (x_6) and usar land but this relationship is not strong because it is significant at 10 per cent level of significance. The other variables *viz.* slope (x_1), ruggedness number (x_2), drainage density (x_3) and quality of water in pH (x_4) have positive correlation with the distribution of usar land is Amethi, but these variables fail to explain the variation in the usar land because their relationship is insignificant.

The values of the different independent variables (x) for usar land (y) indicates a high degree of multi-collinearity among these variables. To remove inconsistencies arising from the multi-collinearity, step-wise regression analysis has been attempted. In this exercise, percentage of usar land is considered as the dependent variable (y) and other natural and human factors as independent variable (x). The inter- correlation matrix has been obtained from step-wise regression analysis. The most dominant variable in step-wise regression analysis in explaining the variability of usar land is the pH value of the soil and Scheduled Caste population. The step-wise regression analysis indicates that 14.4 per cent variation in the proportionate distribution of usar land at village level in Amethi. In the subsequent step variable (x_9) Scheduled Caste population enters and jointly with the variable (x_7) pH value of the soil explains 18.9 per cent total variation in the proportionate distribution of usar land at village level in Amethi. The other variables could not enter in the step-wise regression analysis which indicates that there are other factors that these variables which are responsible for the proportionate distribution of usar land. Thus, among the selected variables pH value of the soil and percentage distribution of Scheduled Caste population are the important variables in explaining the variations in usar land at village level in Amethi Block of Sultanpur District.

Banjar Land

It has been observed that there is positive correlation between the percentage distribution of banjar land (y) and pH value of the soil (x_7) which is significant at 5 per cent level of significance. The high pH value of the soil has high concentrations of salts, which hampers the crop production. The variable (x_9) Scheduled Caste population has positive correlation with the banjar land which is significant at 5 per cent level of significance and the variable (x_8) land concentration has positive correlation with the banjar land. The relationship between banjar land and land concentration is significant at 5 per cent level of significance. The variable quality of water in PH (x_4) and variable (x_3) drainage density have positive correlation with banjar land and their relationships are significant at 10 per cent level of significance. Therefore, these variables are not of much importance in this study because their relationship with variable (y) banjar land is significant at 10 per cent level of significance. It has been recorded that there is positive correlation between the behaviour of water-table (x_5) with banjar land which is insignificant. There is positive correlation with the NPK of the soil (x_6) and banjar land which is significant at 10 per cent level of significance. The slope (x_1) and ruggedness number (x_2) have negative correlation with banjar land but their relationships are insignificant.

The correlation matrix for the various dependent and independent variables shows positive correlations of some of the variables with the proportionate distribution of banjar land. Step-wise regression analysis has been attempted to see as how many of these variables can explain the variation in the distribution of banjar land The variable (x_7) pH value of the soil has been found to be dominant variable in explaining the variation in the banjar land. Step-wise regression analysis explains 23.9 per cent variation of banjar land. The other natural and human factors could not enter in the step-wise regression analysis indicating that these variables fail to explain the variation in the proportionate distribution of banjar land at village level in Amethi Block.

Old Fallow Land

It has been observed that there is a positive correlation between pH of the soil (x_7) and proportionate distribution of old fallow land. The relationship between these variables are significant at 10 per cent level of significance. Variable slope (x_1), ruggedness number (x_2), drainage density (x_3), quality of water in pH (x_4) and behaviour of water-table (x_5) have insignificant correlationship with the percentage distribution of old follow land at village level in Amethi Block.

The calculated correlation matrix indicates the multi-collinearity among the independent indicators. To remove the inconsistencies, the step-wise regression analysis has been done. In the first step of regression analysis variable (x_7) pH value of the soil enters, it explains 11.3 per cent variation in the proportionate distribution of old fallow land. The other variables do not get incorporated in this analysis which indicates that other natural and human factors indicators are incapable of explaining the variations in the proportionate distribution of old fallow land. In spite of these indicators, there are few other factors which are responsible for the uneven distribution of old fallow land at village level in Amethi Block.

Fallow Land

A significant positive correlation has been found between the fallow land and drainage density (x_3). The relationship between these two variables is significant at 5 per cent level of significance. The variable quality of water in pH (x_4) has positive correlation with the proportionate distribution of fallow land which is significant at 5 per cent level of significance. The behaviour of water-table (x_5) has positive correlation with the fallow land which is significant at 10 per cent level of significance. The relationship between the NPK of the soil (x_6) and fallow land is recorded negative and this relationship is significant at 10 per cent level of significance. The pH value of the soil (x_7) and land concentration (x_8) and Scheduled Caste population (x_9) has positive correlation with the proportionate distribution of fallow land. The correlation between X and Y variables are not of much importance, their relationships are significant at 10 per cent level of significance. Slope (x_1) and ruggedness number (x_2) have negative correlation with the fallow land which is insignificant.

The step-wise regression analysis indicates that in the first step variable pH value of the soil (x_7) enters and explains 22.9 per cent variations in the distribution of fallow land. In the second step variable (x_3) drainage density enters jointly with the variable x_7 explains 25.3 per cent total variations in the proportimate distribution of fallow land at village level in Amethi Block. The other variables do not get incorporated in the step-wise regression analysis which indicates that these variables would not explain the variation in the proportionate distribution of fallow land at village level in the Amethi Block.

Other Types of Wastelands

The relationship among these two variables is significant at 10 per cent level of significance. The variable land concentration (x_8) and Scheduled Caste population (x_9) has significant positive correlation with other types of wastelands. The relationship between these variables are significant at 10 per cent level of significance, because the large size landholder do not pay much attention for the other wastelands reclamation and the Scheduled Caste population are oftenly the landless and they have been allotted the other types of wastelands on *patta*, which is not yet reclaimed because of the ignorance and socio-economic backwardness of such *patta* land allottee Scheduled Caste population, that is why such lands are still lying uncultivated and demarcated as other types of wastelands. The cultivators are engaged in cultivating such lands which are profitable and do not pay attention in the cropping of any such land which is expensive in cropping and benefits accrued from the cultivation is lower than the expenditure of cropping. Thus, due to low fertility and productivity of the lands such other types of wastelands are left uncultivated and the cultivators are having traditional farming system and within the limited input, cultivators are unable to cultivate the other types of wastelands. The other variables *viz.* slope (x_1), ruggedness number (x_2), drainage density (x_3), quality of water in pH

(x_4), behaviour of water-table (x_5) and pH value of the soil (x_7) have not much importance in explaining the correlationship between the proportionate distribution of other types of wastelands, their relationship are insignificant. The system of explaining with endogenic variables discussed with the above variables shows the existence of a high degree of multi-collinearity among each other.

The most dominant variable in the first step of regression analysis is (x_8) land concentration which explains 16.9 per cent variations in the proportionate distribution of other types of wastelands at village level in Amethi Block. The other variables could not come in the optimal regression line, indicating their failure to explain the variations in the proportionate distribution of other types of wastelands at village level in Amethi Block.

Wastelands

It has been observed that excess drainage density (x_3) causes much soil erosion, which leads towards land degradation and accumulation of water in the low-lying areas. Thus, high drainage density is associated with high proportionate distribution of wastelands. There is positive correlation between drainage density and the proportion of wastelands which is significant at 1 per cent level of significance. High pH value of the soil has high salinity which causes hindrance in crop production and responsible for the formation of wastelands. It has been proved statistically that the relationship between these variables is significant at 1 per cent level of significance.

The land concentration is found to be high in those villages where percentage distribution of wastelands are also high. It indicates that there is positive correlation between land concentration (x_8) and the proportionate distribution of wastelands. The farmers having large size of landholdings are least interested in maintaining the fertility status of the soil due to high expenditure and very low output from such lands. The cultivators concentrate on good quality lands from which they are sure to obtain good returns. The positive correlation between land concentration and wastelands is significant at 1 per cent level of significance. There is positive correlation between Scheduled Caste population and the wastelands distribution at village level in Amethi Block because the Scheduled Castes are oftenly the landless and as per the Government Policy the landless Scheduled Caste people are allotted land on *patta* which is normally the wastelands. The *patta* allotted wastelands are mostly lying uncultivated due to ignorance and socio-economic backwardness of the Scheduled Caste population because the Scheduled Caste people are resourceless and within the limited resource they are interested for their livelihood and survival of life but not to expend their resources in the cultivation/reclamation of such *patta* allotted wastelands which require high expenditure and very low productivity and least benefit, due to which such *patta* allotted wastelands are lying wastelands. The high prercentage of Scheduled Caste which are mostly landless also indicates the land concentration among very few people. The concentration of land with limited people also proves that the large size landholders are interested to maintain the productive and fertile land but not to invest the finance in such land which requires high expenditure and very low output, due to which high percentage Scheduled Caste population dominated villages are also having very high percentage of wastelands. The positive relationship between Scheduled Caste population (x_9) with wastelands (y) is significant at the 1 per cent level of significance. The variables slope (x_1) and ruggedness number (x_2) are not important in explaining the variations in the proportionate distribution of wastelands because their relationship among these variables with wastelands are insignificant.

It is observed from the step-wise regression analysis that (x_7) pH value of the soil enters in the first step and explains 72.0 per cent variation in the proportionate distribution of wastelands at village level in Amethi Block. In the second step variable (x_8) land concentration jointly with variable (x_7) pH value of the soil explains 79.5 per cent variation in the proportionate distribution of wastelands at village level in Amethi Block. In the third step variable (x_9) Scheduled Caste enters along with pH value of the soil (x_7) and land concentration (x_8) and explains 80.9 per cent total variations in the distribution of wastelands. Lastly, in the fourth step variable (x_5) behaviour of water-table enters jointly with (x_7) pH value of the soil, (x_8) land concentration and (x_9) Scheduled Caste population, which explains 82.8 per cent variations in the proportionate distribution of wastelands at village level in Amethi Block of the District Sultanpur. Thus, the total variations in the proportionate distribution of wastelands explained by selected natural and human factors has been recorded 82.8 per cent. However, the interaction between the set of natural and human factors and the proportionate distribution of wastelands provides a great clue of great significance at 1 per cent level of significance, while the main variables explaining the percentage of wastelands are pH value of the soil, land concentration, Scheduled Caste population and behaviour of water-table.

Thus, it has been observed that man and nature relationship has been shattered because most of the environmental indicators are positively responding to the growth in the wastelands area. The Scheduled Caste population which are mostly landless population are completely dependent of the nature for fuelwood, fodder and food grain etc. are responsible for the wastelands. The increasing population has the major role in the exploitation of natural resources especially the land resources, which is causing natural, ecological, environmental land degradation and crisis in the ecosystem of the area. The severe ecological imbalances has also caused socio-economic crisis in the rural areas.

5. Conclusion

The relationship between man and nature is symbiotic. The combination of human and natural factors plays an important role in the spatial distribution of wastelands and its various categories. The correlation matrix and step-wise regression analysis statistically proves that slope and drainage density are the dominant variables in explaining the variations in wastelogged land at village level in Amethi. The main important variables in explaining the proportionate distribution of usar land are the pH value of the soil and Scheduled Caste population followed by land concentration. The important variable in explaining the distribution of banjar land are the pH value of the soil, followed by Scheduled Caste population. The pH value of the soil, followed by the Scheduled Caste population and land concentration are the major dominant variables in explaining the variations of old fallow land at village level in Amethi Block. The pH value of the soil and drainage density are the main dominant variables in explaining the proportionate distribution of fallow land at village level in Amethi. The land concentration is one of the main dominant variables followed by the Scheduled Caste population and pH value of the soil in explaining the distribution of other types of wastelands.

The main dominant variables in explaining the proportionate distribution of wastelands at village level in Amethi Block are the pH value of the soil, land concentration, Scheduled Caste population and behaviour of water-table. Thus, the variables which features dominating in explaining the variations in wastelands are the natural factors which are pH value of the soil, drainage density, behaviour of the water-table and human factors which comprises mainly the land concentration and Scheduled Caste population. Thus, lastly it can be said that the natural and human factors including some of the other variables are directly or indirectly responsible for the proportionate distribution of wastelands at village level in Amethi Block, District Sultanpur.

4

Salt-Affected Wasteland Soils of Arid Zone and their Development Potential

R.P. Dhir and *A.S. Kolarkar*

Lands comprising sand dunes and sandy hummocky plains, or shallow soils with compact hard pan close to or exposed to the surface, or gravelly lands of skeletal soils on hills or their skirting slopes are the common types of wastelands in the arid zone of western Rajasthan. However, in addition, salt-affected lands with a high amount of soluble salt accumulation in soil profile of otherwise productive soils also form a sizable areas of wastelands in this region. High amounts of salts in soils coupled with aridity of climate and paucity of good quality and quantity of water for their reclamation create exceptionally severe limitations in the development of such lands and make greening such lands a challenging task.

Extent of major salt-affected lands in arid Western Rajasthan

A number of saline depressional basins or "playas" locally called "Rann" are dotted throughout the region and cover vast areas. The notable ones are Pachbhadra, Sanwarla-ka-Rann, Kharia ka Rann, Pokran ka Rann, Lawan-ka-Rann, Thob-rann. The salt content in many of these being very high, the lands are barren. However, at some of these basins, the salts are extracted commercially (Roy and Kolarkar, 1977). Besides these, there are two major tracts in the region where incidence of natural soil salinity assumes an unusual proportion (Dhir *et al.*, 1979) turning them into wastelands of plains. One is in the extreme north-west in the far flood plains of old Ghaggar system covering 0.2 m ha and the other is a large tract in the south-east, identified by Bilara Desuri-Jalor Triangle, covering nearly 0.5 m ha out of which nearly 0.3 m ha is under natural saline wasteland soils (Dhir *et al.*, 1979; Kolarkar *et al.*, 1980). The former has fortunately come under the Indira Gandhi Canal Command System, providing the needed quality and quantity of fresh water either for reclamation for agriculture or for successful tree plantation, depending upon the soil characteristics. The latter tract, however, is faced with dual problems of soil salinity and paucity of water. Groundwater resource is poor and waters are of moderate of high salinity. It is here, therefore, that afforestation of saline wastelands is a challenging task.

Major Land and Soil Characteristics

Salt-affected soils in this part occur on nearly level to gently sloping plain lånds and depressional areas, as irregular shaped patches, varying from 30 to 3,000 ha in area, in association with normal soils. From the pattern of its occurrence and field and laboratory studies of deeper stratas, patchy surface salinity seems to be only a proverbial tip of the iceberg there being far greater mass of salinity in the underlying deep strata and the groundwaters.

In the transitional areas and the so-called relict saline soils (Kolarkar *et al.*, 1980), the mean profile salinity values are mostly 4 to 12 mm hos ECe. But the major areas of salt-affected wasteland soils have much higher salinity; the mean profile ECe being mostly between18 and 65 mm hos (Table 4.1). Therefore, these soils are highly saline indeed. The salinity profile is variable, however, in most of these soils, the highest salt value being observed in surface 10-15 cm, except during a short period of the rainy season. In some situations, there may be another maxima at 80-150 cm soil depth. Sodium chloride and sodium sulphates are by far the dominant salts. The Sodium Absorption Ratio (SAR) of saturation extract is mostly between 35 and 95.

The soils are generally deep and have loam-to-clay loam texture. These are light grey brown to brown and have an angular blocky structure. Often there is dense lime nodular strata at a depth of 90 to

Table 4.1: Salt Composition of some Common Salt-affected Wasteland Soil of Arid Zone

Profile No.	*Depth*	*pH*	*ECe*	*Cations (meq/lit)*				*Anions (meq/lit)*			
				Na	*K*	*Ca*	*Mq*	*Cl*	*HCo_3*	*So_4*	*SAR*
LB 580	0-5	7.8	48.8	0.2	384.2	63.0	57.4	462.4	7.6	35.2	49.4
	5-10	7.6	21.6	0.2	188.4	21.6	23.2	203.6	6.6	22.6	38.9
	10-20	7.8	17.3	0.2	135.5	24.2	18.8	143.6	7.5	24.6	28.9
	20-40	8.0	12.5	0.2	90.5	16.6	18.6	96.2	6.2	22.6	19.1
	40-60	8.2	11.4	0.2	85.1	14.3	13.6	92.2	6.1	18.2	23.0
	60-80	8.1	11.3	0.2	84.8	13.6	14.3	91.4	6.0	19.6	22.1
LB 40	0-10	8.2	48.7	413.0	4.4	35.0	35.0	404.0	2.25	80.1	84.1
	10-30	8.1	49.1	434.7	5.1	28.0	38.0	475.0	1.75	29.0	100.1
	30-50	8.6	33.1	282.6	0.5	24.0	18.0	322.5	1.5	0.18	62.6
	50-75	8.5	39.3	247.8	17.8	29.0	4.0	187.5	2.0	209.1	109.9
	75-95	8.7	33.8	293.4	3.7	10.0	16.0	255.0	1.5	66.6	81.5
PL 22	0-30	8.4	6.8	48.5	0.3	13.2	3.5	57.5	6.0	16.7	16.7
	30-70	8.3	13.6	113.0	0.1	11.5	15.5	112.5	6.5	30.3	30.3
	70-100	8.2	14.3	121.7	0.1	18.7	15.3	115.3	7.0	29.6	29.6
	100-120	8.2	18.3	143.5	0.1	32.9	8.8	145.0	4.0	31.4	31.4
LBS/85	0-5	7.8	132.1	1004.3	0.4	150.0	132.0	1120.0	2.0	164.7	84.5
	5-15	7.6	98.2	692.3	0.4	147.0	135.5	780.0	2.2	192.4	38.2
	15-40	7.0	59.7	421.0	0.2	102.0	69.5	532.5	3.0	57.1	45.4
	40-70	7.3	47.2	347.8	0.1	68.5	53.5	330.0	6.5	133.4	44.5
	70-100	7.3	45.3	347.8	0.2	57.0	47.5	350.0	2.0	99.4	44.4
	100-130	7.3	42.3	301.3	0.2	63.5	53.3	362.5	9.0	46.9	39.3

120 cm. From the fertility point of view, these soils are comparable with adjoining farm land soils, both with respect to major and micro-nutrients (Joshi *et al.*, 1985). In certain situations, highly saline water may be present at depths 3 to 6 metres during the dry period, which may come within 1.5 to 2 m depth in a year of abnormally heavy rains.

Present Land Use and Land Cover

These lands invariably lie as wastelands, often barren. However, despite high salinity at many places, a sparse and patchy cover of some highly salt-tolerant tree and grass species occur in certain pockets. Among tree species *Salvadora oleoides, Capparis decidua, Tamarix articulata* and *Prosopis juliflora* are common while among sedges and grasses, *Capparis rotundus, Sporobolus marginatus, Cynodon dactilon* and *Dactylocterium sindicum* are present. The specific habitats of these are strikingly hummocky or circular mini-hummocks leaving the inter-hummocks bare. Therefore, the present contribution of these lands to the environment and the economy of the tract is negligible indeed.

Improved Management of Salt-affected Lands

Although these soils are deep with better media for deep root growth, have medium to fine soil texture with fairly high water retention capacity to support plants and occur on extensive plain lands facilitating taking up of large-scale plantation programme, the presence of an extra-ordinary concentration of salts in soil profile alone is a very serious limitation for establishment and growth of plantation on these lands. Leaching down the salts out of soil profile, such as has been attempted with a large measure of success elsewhere in the country, is not feasible in this region, because of unavailability of fresh or even moderately saline waters in sufficient quantities. Further, in a number of situations of salt-affected lands, the groundwaters are sufficiently close to the surface, so that a danger of salt aggravation with further rise in groundwater level, and resalinization always persists. Therefore, the approach has not to be of reclaiming and ameliorating the soils but only to create conditions that useful and salt-tolerant vegetation species can establish on such lands and persist despite the hazard. Unlike on normal soils, where there is a wide choice of common, easily available and useful vegetation species to be grown, here on such sites, there is a limited choice of only salt-tolerant species that can be planted. Such plants need to be carefully planted and established on various kinds of salt-affected soils. The growth rate of yield returns of such species on such soil lands may be limited, compared to normal soils, yet the lands need to be treated for their productivity and improving environment rather than allowing to remain waste.

Selection of the tolerant species is one ingredient of the technology and establishing of these two with careful management to prevent disturbance and destruction by animals and man can lead to a success of plantation and also to amelioration of the condition of such lands.

Choice of Salt-tolerant Species

Based on the mean profile salinity, an appropriate plant species should be chosen. In common text literature a long list of salt-tolerant forage and tree species can be obtained but all may not suit to the arid environment. Based on the results obtained and experiences gained in the region at CAZRI, Jodhpur, a general relative salt tolerance of the tree and forage species suitable for aid zone is summed up in Tables 4.2 and 4.3 respectively.

Table 4.2 : Relative Salt Tolerance of Different Tree Species

Tolerant	*Moderately tolerant*
Eucalyptus camaldulensis	Eucalyptus hybrid
Casuarina torulosa	Leucauna leucocephala
Tamarix articulata	Colophospermum mopane
Tamarix sp	
Prasopis juliflora	
P. Challensis	
Capparis aphylla	
Atriplex vesicaria	
Acacia nilotica	
A. ropers	
Alianthus altissima	
Acacia aneura	
Chenopodium species	
Acacia tortilis	

Another problem at most of the places in the region is the scarcity of fresh water even to raise enough number of plant saplings in the nursery. Groundwaters available are generally saline to highly saline. Studies conducted at CAZRI (Jain *et al.*, 1982, 1983) indicate that with water of as high salinity as 9,000 micromhos EC, saplings of species like *P. juliflora*, *Tamarix articulata* can be raised for as long as months while with waters of 6,000 micromhos EC, *Eucalyptus hybrid*, *Dichrostachys glomerata*, *Acacia aneura* can be raised for as long as 6 to 9 months. In addition to these species, *Cassia siamea* and *Calophosphermum mopane* can also be raised with irrigation with high saline waters at least for 3 months. Irrigation in the early stage is essential in this region and if such plants are transplanted on normal (non-saline) wastelands, they can easily be established with irrigations with saline waters which is locally available.

Table 4.3 : Different salt tolerant forage species

Tolerant	*Moderately tolerant*	*Semi-tolerant*
Sporobolus marginatus*	Cenchrus setigerus	Heteropogon contortus*
Eleusine compressa	Aristida sp.	Zizyphus nummularia*
Fagonia cretica	Eremopogon foveolatus*	Tribulus terrestris
Tavernia cunefolia	Lepidagathis sp.	Brachiaria ramosa
Barleria acanthoides	Tephrosia purpurea	
	Digitaria adscendens	
	Indigofera oblongifolia	
	Sehima nervosum*	
	Convolvulus arvensis	
	Boerhavia diffusa	

* Important species of high forage value.

Planting Techniques

Practical method of vegetating salt-affected lands is to create pockets of low salinity in immediate vicinity of the plant roots. This can be achieved in two ways. First by giving the land ridge and furrow configuration. The ridges may be 0.8 to 1 m high in relation to the bottom of the furrow. Their adjustment is purposely kept in east-west direction so that with redistribution of salinity the ridges are relatively free of salinity. The preferred adjustment helps in concentrating the leftover salinity on the southern face on the ridge because of greater evaporation.

The other alternative, particularly for widely spaced plants, is the pit method of planting. In this method, the pits are dug and filled with externally brought non-saline soil mixed with sheep pan or farm yard manure. Though in course of time, the soil gets salinized but by that time the plants are already established and get adjusted to such soil environment.

Watering Plants

Because of the fact that rainfall is low, irregular or erratic in this region, and evapotranspiration is very high, watering of the species is not only necessary in the nursery stage alone but also after transplanting, at least for a period of one year. Waters of low to moderate salinity (2,000 to 6,000 EC) that are available in the region can be used for tolerant species, and the irrigation schedule has to be little more to leach down the accumulated salts. There may not be need of irrigation during the next monsoon, once the plant is established.

Summary

Largely, the salt-affected wastelands occur, particularly concentrated in two tracts, namely, in the north-west and in the south-east part of western arid Rajasthan. High amounts of soluble salts in the soils, coupled with high aridity of climate, create exceptionally severe limitations on the soils to grow vegetation. Except for this limitation, these are generally plain lands, with deep medium to fine textured soils with a fairly high moisture retention capacity and fertility, comparable with any other farmland soil. As such they have otherwise high production potential. The north-west tract has fortunately come under the Indira Gandhi Canal System but in the latter tract, leaching of salts does not appear a possibility because of paucity of fresh waters. Nevertheless, considerable scope seems to exist of raising productivity of such saline wastelands through plantation of salt-tolerant forage and, tree species, with suitable management practices to create pockets of low salinity. There are a number of such species available, which once established can withstand both the adverse conditions of salinity and aridity.

REFERENCES

Dhir, R.P., Singh, N. and Sharma, B.K. (1979). *Ann. Arid Zone* 10 (2&3). 27-34.

Kolarker, A.S., Dhir, R.P. and Singh, N. (1980). *J. Indian Soc. Photo Inter and Remote Sensing* 8 (1), 31-35.

Jain, B. L. and Muthana, K.D. (1982). *My Forest* 18 (4), 175-180.

Jain, B.L., Goyal, R.S. and Muthana, K.D. (1983). *Ann. Arid Zone* 22(3), 233-238.

Joshi, D.C., Singh, N. and Kolarkar, A.S. (1984). *J. Indian Soc. Soil Sci.* (Communicated).

Ray, B.B. and Kolarkar, A.S. (1977). *Nature and Resources of Rajasthan* (ed. M.L. Roonwal), Jodhpur University, Jodhpur, Vol. 2, 1011-21.

5

Wastelands Development : A Case Study for *Prosopis Juliflora*

D. P. S. Verma

Wasteland is defined as that land which is presently lying unutilized due to certain constraints. It comprises cultivable land and uncultivable land. Dr. Sekhar of Ford Foundation considers any land producing less than 20 per cent of its potential as wasteland. According to the National Wastelands Development Board, "Any land if its productivity is less than what it ought to be with reference to the availability of soil nutrients and water should be considered as wasteland". Going by the latter definition, most of the Indian land mass will fall under the category of wasteland; land wasted by human race through its various acts of omission and commission. Disforestation in the most part and excessive use of water in some have rendered large tracts of the country's land unproductive to a stage that retrieval seems an uphill task if not impossible.

The *Handbook on Agriculture-1980* recognizes the following land uses in India:

Sr. No.	*Land Uses*	*Area under various land uses in m.ha*	*% of total area*
1.	Net area sown	143	43.47
2.	Area under forests	67	20.36
3.	Under non-agriculture use	18	5.47
4.	Pastures and groves	16	4.86
5.	Cultivable waste	17	5.17
6.	Fallow land	22	6.69
7.	Barren and uncultivable waste	22	6.69
8.	Area for which no return exists	24	7.29
	Total	329	100.00

Earlier Five Year Plans laid an undue emphasis on extension of land under agriculture and a fairly large chunk of good forest was lost under the Grow More Food Campaign. The situation worsened in the hilly terrain where luxuriant forests were cut and land went under submergence for hydro-electric-cum-irrigation projects. A large forest area was also utilized for re-settlement of the project affected villagers. The State of Gujarat, having only 10 per cent (1.96 m.ha) land area under forest, lost over 57,000 ha during the period 1960-80. It is true that due to increase in the irrigation potential and intensive agriculture, today we are not only self-sufficient in food production but are in a position to export large quantities of wheat and soyabean. But, we are very much short of firewood and other forest products. Today, a stage is reached where we have enough of food but not adequate firewood to cook the food with. Time has come when some rethinking is necessary regarding extension of agriculture to unremunerative or less productive agricultural lands. All such lands must be diverted to tree cropping. It may be necessary to redefine the frontiers of land use, according to its capability. The best land use is said to be one user which the land produces the most and deteriorates the least. The land capability should mean not only the present capability of land to produce, but also to retain the potentiality under a given land use in perpetuity. The land must be used according to its capability and treated according to its needs.

During the past one decade or so, the area under agriculture has remained nearly static — an indication of non-availability of more arable land. This fact, seen in the context of ever-rising human and cattle population, throws some light on the condition of remaining land which is obviously so bad that it cannot be economically brought under the plough. The factors responsible for degradation are many—infertility, salinity, waterlogging, erosion, etc. This also indicates that, by and large areas left unutilized are either highly refractory or they have been subjected to incessant blows of mismanagement and turned into saline, eroded or barren lands. The pasture and grazing lands, which constitute about 5 per cent of the total land area are the most degraded. They are maintained in that state with the ever-increasing biotic pressures. Every twig jutting out its head during the rainy season is either ripped or trampled by the cattle, or is cut and used by the people to heat their hearths. The pasture lands wear a naked, neglected and weary look, year after year.

The "Wastelands Development", a document circulated in early 1986 by the National Wastelands Development Board (NWDB) concedes that around 175 m.ha is degraded land of various descriptions; 40 m.ha in agriculture; 30 m.ha in forestry and over 100 m.ha of pasture and grazing lands, the hot desert of Thar and the cold desert in the northern parts. Going by the kind of degradation, the wind and water eroded wastelands are approximately 13 m.ha and 73.5 m.ha respectively. Wastelands due to salinity and alkalinity account for over 7.0 m.ha Although, the records show about 77 m.ha as cultivable waste, fallow land, barren land, pastures and groves, it is difficult to presume that all or even a major part of this land would be available for tree planting activity. A good deal is under encroachment by vested interests enjoying some degree of political and administrative immunity in one form or the other. A fair proportion is practically uncultivable.

Jonathan Swift was more than candid when he wrote in *Gulliver's Travels* that "whoever could make two ears of corn or two blades of grass grow upon a spot of ground where only one grew before, would deserve better of mankind than the whole race of politicians but together". Growing a blade of grass where none grew before, perhaps needs a more eloquent description and demands a really deft human hand. The job of wastelands development is, thus, both difficult and demanding.

Wastelands suffer from one or more of the following infirmities:

1. Steepness of terrain, by and large natural;
2. Low soil depth and/or stony structure;
3. Poor in humus and nutrients;
4. Salinity and/or alkalinity;
5. Low and/or badly distributed rainfall, leading to poor moisture status;
6. Too low or too high temperatures, making cold or hot deserts;
7. Fast wind causing severe erosion;

8. Too much water coupled with poor drainage, leading to waterlogging;
9. Heavy biotic pressure;
10. Local customs such as shifting cultivation, followed by brush wood burning, etc; and
11. Abuses due to illiteracy, wrong notions, greed, etc.

The above list precludes such less, limiting and local factors as frost, hailstorm, salt-laden winds, fires, etc. Many of the factors are bound to reduce the fertility of land specially when less than 2 per cent of the world's forest area has to meet the needs of nearly 15 per cent of its population. This is compounded by the rising standard of living of the masses, irregular distribution of forests and low productivity (0.53 M^3 per ha of our forests against the world average of over 2.1 M^3 per ha). Further, we have lost about 7 per cent of our forest area in a comparatively short span of three decades. On the other hand, over 77 m.ha (175 m.ha according to NWDB) is wastelands. Add to this, the factor of acute poverty, quite a large population of the country is either unemployed or it is underemployed.

Thus, while on the one hand, scarce land resource is lying unutilized, on the other millions of poor people of the country go without bread due to non-availability of opportunities for employment. In other words, we have the resource to use and develop and also the unemployed manpower to work on this resource. It is due to this reason that every sensible man in the country is talking to wed these two massive resources for the over-all improvement of the country. It was under these circumstances that our dear Prime Minister in his broadcast to the nation on January 5, 1985, gave a clarion call to undertake a massive afforestation programme of planting fuel and fodder trees and develop five million hectares of wasteland every year in order to avert a major ecological and socio-economic crisis which is looming large on our heads and likely to assume ugly proportions of a national catastrophe. The Prime Minister's call has taken the nation by storm and no wonder, the country is gripped with the fever of finding ways and means to fulfil the challenge thrown by the Prime Minister.

Several issues are involved in carrying out this gagantic task.

Managerial, financial, social, legal and technological, such as identifying the wastelands, motivating and involving people on a massive scale, involving N.G.Os, finding financial resources for the programme, selecting proper species and developing methodology of operations are some of the basic issues involved in the programme. The latter two are the tools to be handled by technology. The Prime Minister himself has indicated the broad parameters for choice of species inasmuch as the species to be selected should be in a position to provide fuel and fodder to the Indian masses. Though, broad outlines for choice of species have been laid down, their final selection has to be done by the experts, depending upon the site conditions, climatic limitations and above all the need of the people. Within the broad framework of policy, it is the site which would decide the choice of species. Fortunately for us, we have a very broad spectrum of multipurpose tree species available in the country to choose from. *Prosopis juliflora* fulfils this in more respects than any other single species for most of the sites termed wasteland simply due to the reason that the areas to be tackled are, by and large, very hostile sites and there is heavy biotic pressure on them by way of cutting for fuel and grazing for cattle, goat and sheep.

Prosopis juliflora, which is aptly known as Gando-baval (mad-Acacia) in Gujarat, is native of Central America and the West Indies. Due to its hardiness, wide adaptability to various adverse sites, non-acceptability of its leaves and twigs by cattle as grazing material, profuse seeding followed by heavy browzing of pods by sheep and goat thereby ensuring its dissemination and propagation, and many other favourable characters. *Prosopis* is amongst the most widely distributed trees in the world. At present, it is spread over the continents of South America, Africa and Asia. It is either growing wild or is cultivated on a large-scale in north-eastern Brazil, Peru, Sudan, several Sohelian countries, South Africa, India and Pakistan. As in India and in some other countries, it grows so profusely that it is considered a weed.

Over 40 species of *Prosopis* are identified and reported so far in the world. Some of them are aggressive bushes, while there are a few which produce useful forage and wood. All of them do grow under some degree of adversity of one kind or the other like climatic, edaphic or biotic. *Prosopis juliflora* has proved the most aggressive in our country and it is thriving more luxuriantly than many local and indigenous species. Trials with various species of *Prosopis* at Bhavnagar have shown that few of them hold good promise for our country. Prominent amongst them are *Prosopis velutina* (provenance 475)—quite fast growing and drought resistant; *Prosopis alba*—a fast-growing tree with a high content of sucrose and proteins in its pods which are valuable as cattle feed; *Prosopis glandulosa* and *Prosopis pallida* are very agressive and fast growing in their native land but in comparison their rate of growth is rather slow at Bhavnagar though the general state of health of the plants is good; *Prosopis velutina* (provenance 457) can withstand very high salinity of water; *Prosopis alba* (provenance 388) is a thornless species and has shown a very promising rate of growth under Bhavnagar conditions. Its leaves and tender branches are relished by cattle unlike most of the other species of *Prosopis* genus. It is a good species for areas suffering from fodder crises.

This wonder child of nature (*Prosopis juliflora*), introduced in India somewhere in 1876 due to the efforts of Lt. Col. R. H. Beddome, Conservator of Forests of Northern Circle of Madras Presidency, was first tried at Camalapur in Cuddapoh district of present Andhra Pradesh. This delicate and shy looking plant took to India as fish takes to water, and never looked back right from its first day of introduction and slowly established its domain over arid, sterile, saline, degraded and sandy sites. Such hostile sites, which had no suitable tree to clothe them, found a good shady tree with evergreen foliage in *Prosopis* and it soon became a darling of millions of people who live in the vicinity sites throughout the length and breadth of the country at times with the efforts of foresters and more often without any attempt whatsoever. Though exotic, *Prosopis juliflora* has made India its home and has spread rapidly, got established, acclimatized and holds great promise for the hostile sites particularly in arid and semi-arid saline wastelands.

Prosopis juliflora can claim for itself first preference to establish itself as one of the major species for wasteland development due to the following unsurpassable qualities:

1. Source of Fuel

The most important use of *Prosopis juliflora* is as fuel. Its capacity to produce fuelwood over a short period specially on hostile land like wastelands is unsurpassable. It produces an excellent firewood for cooking stoves, and for heating in rural areas. The calorific value of *Prosopis* wood is well comparable with other trees grown in semi-arid areas which is clear from Table 5.1.

Apart from the calorific value, the wood of *Prosopis juliflora* produces very little quantity of ash (2.9%) and yields excellent charcoal of gun-powder quality which is termed "Arthracits coal" for its high calorific value. The wood contains about 40 per cent lignin which imparts high heat of combustion to the wood. The root wood is very hard and a top class firewood with insignificant sap wood and is highly valued for household or restaurant cooking. The fuelwood burns with little smoke and is much sought after for barbecues as it imparts a delicious flavour to the meat. As recent as World War II, the firewood was used to power steam engines and industrial boilers in Argentina in place of coal.

Table 5.1 : Comparative calorific value of some of the important fuelwood species of semi-arid areas

Sr. No.	*Name of Species*	*Calorific Value Cal./Kg.*
1.	Acacia auriculiformis	4800
2.	Acacia nilotica (sap-wood)	4800
	Acacia nilotica (heartwood)	4950
3.	Acacia leucophloea	4886
4.	Prosopis juliflora	4800
5.	Acacia tortilis	4360

2. An Excellent Producer of Biomass

Prosopis is an excellent producer of biomass. It ranks first amongst the high biomass producing native trees of arid and semi-arid areas of India. An experiment to determine biomass productivity of important native trees of semi-arid areas was undertaken by the Forest Department of Gujarat State during 1979 under rainfed conditions at Gandhinagar. In all, seven local fast growing, fuelwood species, were planted under identical field conditions and their total biomass production was determined periodically. The work of recording of data of biomass production was carried out at regular interval by Gurumurti of Plant-physiology Branch of Forest Research Institute, Dehra Dun. The different species produced, the following quantities of biomass at the end of 5th year (Table 5.2).

Table 5.2 : Biomass production by different species at the end of fifth year

Sr. No.	*Name of Species*	*Biomass Production in tons/ha.*		
		Utilizable	*Non-utilizable*	*Total*
1.	Albizia lebbeck	59.6	6.7	66.3
2.	Dalbergia sissoo	75.5	6.5	82.0
3.	Eucalyptus hybrid	76.1	7.3	83.4
4.	Cassia siamea	77.5	7.9	85.4
5.	Acacia nilotica	110.1	3.4	113.5
6.	Acacia tortilis	113.9	3.0	116.9
7.	Prosopis juliflora	158.3	8.9	167.2

The figures in Table 5.2 conclusively prove that *Prosopis juliflora* is miles ahead in biomass production when compared to its other associates which are normally grown in dry zone areas. This comparison is shown in a bar diagramme as under:

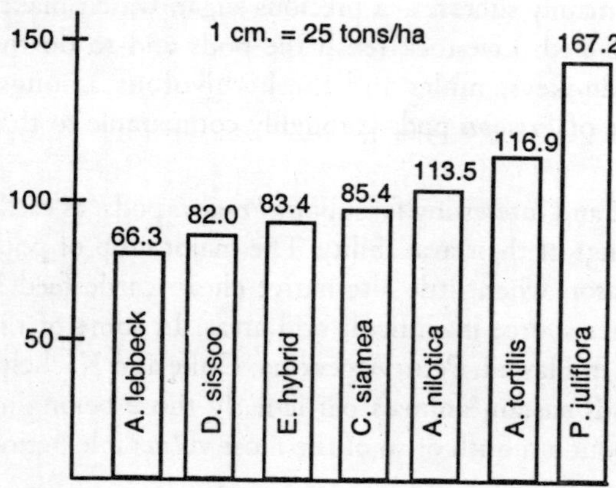

3. Source of Cattle Feed

The most negative single factor against *Prosopis juliflora* as a prospective species for wasteland development is its incapability to provide leaf fodder for cattle and presence of thorns which makes working with it difficult. But one thing should not be lost sight of, that raising of trees is becoming more and more difficult under heavy biotic pressure of a large number of cattle let loose to roam about. Though *Prosopis* leaves do not attract animals for fodder, this is more than compensated by its pods. It is reported that the *Prosopis* pods were used as food by human beings in the new world and even today they are a source of cheap supply of carbobydrates and proteins for many dwellers of deserts in North America. The pods are produced within a very short period so much so that a two-three-year-old tree starts producing pods and profuse production starts from the fourth year onwards. Another interesting and favourable feature of pod production is that they are produced twice a year and do not split on drying like other legumes. Consequently, neither pulp nor seeds are lost. This enables the locals either to use them directly or to collect them for their use at a later convenient time. Very few other legumes can compete with *Prosopis* in this respect. This feed can be stored over long periods for use during the lean months of the year at later stage without any deterioration in the quality of pods.

The pods yield sweet yellow and sticky pulp within which lie the seeds. The pulp is masocarp and its weight varies considerably but experience has shown that it makes about 40 per cent of the weight of dry pods. The pulp is mainly sucrose – a precious sugar, which makes the pod an ideal cattle feed. Livestock relish the pods and so do the sheep, goats, horses, donkeys, mules and the herbivorous amongst wildlife. The feed value of *Prosopis* pods is roughly comparable to that of barley or corn.

Another important and interesting fact about *Prosopis* pods, as cattle feed, is about the timings of their availability. The major crop of pods is produced in dry season when little alternative cheap cattle feed is available from any other source in difficult arid areas. In some of the most hostile arid areas in Hawaii, Peru, Argentina, Chile and Kachchh (Gujarat) in India, pods sustain animals particularly those belonging to poor villagers for about a month or so of the most vulnerable period of the year.

Prosopis seeds are valuable as cattle feed from nutrition point of view also. They contain about 30 to 40 per cent proteins and about

7-8 per cent of oil. The only limiting factor is that to make use of these nutrients by animals, the seed or for that matter the entire pods are required to be crushed or ground otherwise the seed, due to presence of hard coat, would pass through the digestive tract unutilized. Another constraint for use of seed is its difficulty in grinding due to presence of sticky pulp surrounding it as this makes the grinding difficult.

There used to be a feeling amongst villagers that the *Prosopis juliflora* pods are toxic as cattle feed but this is not borne out by actual experience. Pods of some of the other species of *Prosopis*, like *P. pallida*, *P. glandulosis*, etc., are reported to be having toxicity in them and it affects the cattle health and causes "jaw and tongue" trouble, loss of weight and difficulty in chewing but no such effect is reported from pods of *Prosopis juliflora* in India. Kargarrd and Van Des Maruse (1976) had reported that *Prosopis juliflora* growing in South Africa has proved to provide a fairly well balanced, highly digestive diet for sheep, without any ill-effect even when they were kept exclusively on these pods for prolonged periods.

4. Source of Honey

Prosopis vegetation has proved as a favourable home for honey bees. Wherever *Prosopis* has invaded the land, its first effect has been manifold increase in the production of honey in the area. Hawaii, which was not on the honey map of the world, became a leading honey producing country of the world after introduction of *Prosopis*. Even in India, *Prosopis* has lately been recognized as one of the majcr honey- producing tree species. In Gujarat, about one lakh kilograms of honey is coming out annually from the *Prosopis* infected district Kachchh, with a potentiality to increase it manifold. The nectar gathered from *Prosopis* trees by honey bees is reported to be of a superior quality with an attractive flavour.

5. Timber

Prosopis is a medium sized tree with a short trunk and spreading branches. It is, therefore, capable of producing only small size timber specially when grown on hostile sites, which we are likely to come across among the wastelands. Even the small-length timber is usually not straight and it has little use as lumber. The timber is crooked and more often than not hollow. But this small-size timber has certain positive qualities. It has striking grain; it takes very good polish; it is long-lasting in uses involving contact with ground. Wherever small size can be used, it can replace Khair wood – the famous axle wood of India. Due to some of its qualities specially of grain and ability to take good polish, *Prosopis* timber has been equated with Walnut, Rose wood or Mahogony in certain parts of the world (Dobio J. J., 1943, *The Conquering Mesquite*; *National History Museum*, 51, 208-217). The *Prosopis* timber has been used in certain parts of the world for varied purposes like parquet flooring, furniture and turnery items. Blocks of *Prosopis* were first used as street paving in Santonio, Texas. Moreover, as recently as about 30 years back, major avenues of Buenos Aires, Argentina, were "cobble stoned" with cubes of *Prosopis* wood not on sand and coated lightly with tar. The resulting road surface was smooth, noise absorbent and this surface lasted for 10-15 years (*Resource for Future*, National Academy for Sciences, Washington DC., 1979, p. 155). In India, *Prosopis* timber has hitherto found an accepted role only as fence posts and small pillings.

6. Source of Nitrogen

Prosopis is a legume and like any other legume; it fixes atmospheric nitrogen–an essential element needed for growth and development of plants and in which Indian soils are acceptedly poor. This ensures growth and survival of *Prosopis* on poorest of the soils in semi-arid to arid lands that would otherwise remain economically unharnessed. This makes *Prosopis* one of the important prospective tree species with high potentiality for wasteland development. It is also reported (Falpler and Maines, 1977) that soil beneath *Prosopis* trees can be three times richer in nitrogen than soil more distant front the plant. It is due to this reason that forage grass is reported to be growing well in combination with *Prosopis* as the nitrogen and shade provided by *Prosopis* are often evident in the colour and vigour of grasses growing nearby.

7. Drought Resistant Species

Prosopis, though it loves moisture, can grow well in low rainfall areas also. *Prosopis* usually needs an average annual rainfall of about 500 mm (20 in.), but it can establish itself quite well even in areas receiving much less and erratic rainfall. It is reported establisping itself in areas having as low annual rainfall as 3 inch. It has a spreading network of lateral roots that penetrate down to 10 in. depth in search of sub-soil moisture. The roots of one of the sister species of *Prosopis juliflora* namely, *P. volutina* were recovered from an open pit copper mine at a depth of 53 m. in Turson, Arizona. Due to an excellent root system, *Prosopis* is highly drought resistant and is well adapted to the heat of dry regions.

The foregoing paras show that *Prosopis* is a source of food, fodder, fuel wood, shade and shelter in warm deserts of the world like North America, and other similar parts of the world including India. In certain parts of the world, like Atacama and Gram-chaco deserts of South America, where *Prosopis* grows abundantly in nature, it was considered of great value as timber and other products for which no other source was available. Subsequently, natural stands were logged out. *Prosopis* also easily withstands protracted droughts and can be relied on to survive and even produce a crop of pods during a drought year. In the arid parts of India, thousands of villagers, particularly the less privileged and the poor, look up to *Prosopis* as their saviour because it provides them fodder, shade, fuel and source of employment when everything else has vanished in years of long droughts.

Prosopis can stand well to hostile sites of wastelands. It does well even on nutrient deficient poor soils. It can grow well in barren wasteland if it has some depth and the sub-terranean water is available. It also thrives well in light sand or rocky soils. It can also tolerate salts to a great extent, and responds well to the underground saline eater having salinity of 3,000 to 5,000 ppm. Such land is otherwise useless for most of the crops in arid zones. Thus, *Prosopis juliflora*, with all its qualities, is an excellent plant for problem sites in arid areas. It is an excellent species for erosion control, for stabilizing shifting desert or coastal sand dunes. It is also an excellent tree for wind-breaks, shelter belts and for reforesting wasteland in arid and semi-arid areas of the country.

8. An Excellent Coppicing Tree

Prosopis can be propagated very easily by seeds, through suckers, and from root and shoot cuttings. It is an excellent coppicer. Some compare it with God the Almighty. "The more you hurt (cut/fell) it, the more it gives". When livestock feed on whole pod or ungroundseed, the seed passes intact and undigested through the digestive tract of the animal and gets disseminated from place to place by the wandering livestock. The undigested seed with slightly softened coat germinates profusely when lodged in the ground. This quality coupled with its capacity to withstand adverse climatic conditions and heavy grazing has given it the stigma of aggressiveness and spreading nature.

9. Species Undamaged by Enemies of Vegetation

The two main factors which have reduced tree cover in the country are uncontrolled grazing and uncontrolled removal of vegetation for fuel. *Prosopis* can stand up to both these enemies of vegetation and would probably defeat them in a fair war though, at times, it may look losing the local battles.

Economics

A few enterprising farmers have tried *Prosopis juliflora* in commercial plantations on their problematic sites. Shri Kalidas Patel of Vatwa (Ahmedabad) raised *Prosopis juliflora* under rainfed conditions on 24 ha. piece of his land in Adarsh farm of Vatva (Ahmedabad), raised it under rainfed conditions on 24 hectares of land which has pH value of 9.0 which receives annual rainfall of about 750 mm. Input like chemical fertilizers, pesticides and farmyard manure were applied periodically. The crop was harvested at the end of five years. Total expenditure during the five years was Rs. 8,300 per ha. The returns at the end of this period per ha were—pods about 2.5 tonnes values at Rs. 1,250 and firewood 80 tonnes valued at Rs. 24,000. The total returns worked out to Rs. 25,250. Thus, the net returns were Rs. 16,950, i.e. Rs. 17,000 per ha for five years or Rs. 3,390 per ha/year.

Limitations

While Prosopis is an excellent tree species for the hostile sites of arid and semi-arid areas, it has certain limitations. The first and foremost limitation is that it is essentially a tree for arid areas and hostile sites and not quite desirable for good sites with better moisture availability where other valuable trees can be grown. In such areas, there is always the fear of its ousting other valuable species and even encroaching on agricultural fields due to its aggressiveness and spreading nature. It prefers soft soils with some depth and is not happy in rocky, gravelly or pure murrumy and stoney soils. It does not provide grazing which is both in its favour and against it.

Acknowledgement

I am very thankful to Shri M. S. Khanchandani, I.F.S. (Retd.), for kindly going through the draft and making valuable suggestions in the same. I am also thankful to Shri Ashok Kumar, I.F.S., Conservator of Forests, for making valuable suggestions for rearranging the material of the original draft.

REFERENCES

Agro Forestry and Wastelands in India, A paper by V.P. Agarwala, IFS (Retd.), Adviser, Society for Promotion of Wastelands Development.

Agro Forestry for Millions by Soanker Rangnathan, 1979.

"Firewood Farming on Degraded Lands" by A.N. Chaturvedi, *U.P. Bulletin* No. 50, 1985.

Resources for Future, National Academy of Sciences, Washington, DC.

Wastelands Development, A Document of Wastelands Development Board Circulated in early 1986.

Wastelands—Challenge and Response, A Document of N.W.D.B., 1986.

6

Wasteland Development in Haryana

V.N.K. Pillai

Post-independent India made rapid strides in agricultural and industrial development. The improvement in health services and general educational levels has resulted in increase in the average life span of an Indian. However, the rapid increase in the human and cattle population has unprecedented pressures on the limited land resources. About 4.13 million hectares of forest area has been deforested, for settling displaced persons, extension of agriculture, setting up of hydro-electric and irrigation projects, industries, mining, road construction and a number of other non-forestry uses. Extension of irrigation facilities to new areas resulted in utilization of village common lands, Panchayat lands, etc. for agriculture. These areas used to meet the grazing requirements of the cattle population in the villages. Increase in the number of productive cattle put great pressure on the existing land resources, resulting in accelerated erosion, creating the vicious circle of poverty and erosion.

The Government of India has become alive to the problem of the rehabilitation of wastelands, and embarked upon an ambitious task of wasteland development at the rate of 5 million hectares per year.

Wasteland

It has been estimated that a total of 175 million hectares of area exists as wasteland in the country. The estimation mainly done through satellite imagery is disputable. However, there is no doubt about the existence of wastelands in the form of sand dunes, ravines, usar lands, degraded forests and wasted agricultural lands.

Origin of Wasteland

Wasteland occurs naturally due to geological factors or is the creation of human action.

A. Natural Wastelands

(i) Rocky out crops
(ii) Glacial areas and snow-clad areas
(iii) Sand dunes and sandy wastes
(iv) Saline and alkali lands

B. Man-made Wastelands

(i) Eroded lands and ravines
(ii) Waterlogged areas
(iii) Lateritic areas
(iv) Mined areas
(v) Degraded forest areas

(a) *Rocky out crops* and glacial/snow-clad areas are created due to the geological upheavals, lava flow, etc. Extensive rocky out crops exist in the Deccan plateau, Eastern and Western ghats, the Aravallis and Vindhyan Ranges. Glacial areas and snow-clad lands exist in the Alpine zones in the Himalayas. Sand dunes and sandy plains are also formed due to geological changes, but often the destruction of vegetation in such areas results in accelerated wind erosion and problem of moving sands into fertile areas, habitations, etc. *Saline and alkali lands* occur naturally due to lack of precipitation, imperfect drainage, or inundation from sea. In India, many areas on the West Coast and Rann of Cutch suffer from salinity due to inundation. In the Indo-

Gangetic plains, states of Punjab, Haryana and Uttar Pradesh contain extensive Usar lands. The problem can be man-made also, caused by imperfect irrigation in dry areas. More often, such areas contain calcium-carbonate nodules as a pan under the soil, at varying heights.

(b) *Erosion* caused by water gives rise to unproductive lands due to removal of fertile top soil, and rendering the land unfit for agriculture. Erosion is often accelerated by excessive grazing and consequent compaction of soil. Excessive water erosion gives rise to ravine formation, as existing in the Chhamb Catchment, foothills of the Siwaliks, Western and Eastern Coasts.

(c) *Waterlogging* for a long period in a year is often caused by imperfect drainage, caused by the presence of pan formations under ground, seepage from canals, or stiff clayey soils. Waterlogging is also caused by sea water in coastal areas where the lands are below the mean sea level, as in parts of the West Coast. Removal of vegetation in tropical forests also causes formation of Swamps. Waterlogging has rendered large extents of areas in canal irrigated areas unproductive.

(d) *Laterite* formation is a phenomenon found in humid areas where tropical forests are felled. In such areas, the presence of vegetation and addition of humus maintain the chemical balance of the soil. Once the vegetative cover is removed, and top soil washed off, oxides and hydroxides of aluminium and iron get exposed and hard laterites are formed.

(e) *Mining* for building materials, metallic ores, coal, lime stone and brick-making has rendered extensive areas unproductive. Removal of surface soil, formation of large pits and irregular nature of mining cause difficulties in rehabilitating such areas.

(f) *Degradation of forest lands* and other tree lands is caused by a number of factors. Fuelwood collection and uncontrol grazing are the most important ones in the case of forest lands located near habitations. Increase in human and cattle population results in increased demand for food and fodder.

> Absence of alternative fuel, or more often the lack of resource to buy alternative fuel, results in terms to forests for free fuelwood. A large number of rural people are compelled to collect wood and sell in the towns for a living. Absence of employment opportunities in the village, and subsistence agriculture have resulted in conversion of extensive tree lands into wastelands.

Situation in Haryana

Total geographical area of Haryana is 44,212 sq. km. The forest area is only 3.8 per cent of the total area. The National Waste Land Development Board has estimated about 24.4 lakh hectare area as wastelands in Haryana which is about 55 per cent of the total geographical area. Total cropped area of the State is 86 lakh hectares and, therefore, the estimation of wastelands has been on the higher side.

An estimate by the Forest Department gives the following figures:

Degraded forest lands	51,403 ha
Community and panchayat lands	38,654 ha
Agricultural wastelands	30,000 ha
Institutional lands	2,000 ha
Total	1,22,057 ha

Degraded forest lands are mainly located in the Siwalik foothills in the north and in the Aravallis in the south. The latter are mostly owned by private individuals or communities. A large part of this area is rocky and hardly fit to sustain much vegetation due to human action and grazing.

Large parts of the Panchayat lands in Central Haryana are salt-affected areas and alkali lands, waterlogging is common in such lands. Similarly, the agricultural wastelands are also largely salt-affected/alkali lands or sand dunes.

Problematic sites

The following types of problematic sites exist in Haryana:

(1) Degraded forests in the Siwalik foothills, mostly eroded or with ravines with vary little top soil.

(2) Degraded forests/community lands in the Aravallis. The area falls in a semi-arid belt, and growing trees is a problem due to the climatic conditions, lack of soil and multiplicity of ownership.

(3) Saline and alkali lands, mostly occurring in Central and South Haryana in the Districts of Kurukshetra, Karnal, Sonepat, parts of Jind, Hisar and Rohtak.

(4) Sandy plains and sand dunes occurring in the south and south-western district of Mahendragarh, Bhiwani, Hisar, Sirsa and parts of Rohtak.

(5) Waterlogged areas mostly found in central Haryana and along roads, canals and railway lines. Presence of *Kankar pans* (calcium carbonate nodules in compact form) or stiff clay soils inhibit drainage and cause waterlogging.

Methods of Treatment

Foothills of Siwaliks

In these areas, top soil being eroded, soil pH is often more than eight. The compact nature of soil, undulations and ravine formation do not allow moisture to be retained.

Methods practised are sowing/planting in interrupted trenches and mounds, so that precipitation is intercepted and retained in the area. Species used are *Acacia catechu* (Khair), *Lucaena lucocephala* (Subabul), *Acacia nilotica* (Kikar-Babul), *Emblica officianalis* (Amla), *Dalbergia Sissoo* (Shisham), *Holoptelia integrifolia* (Papri), *Toona coliata* (Toon). *Teriminalia alata* (Sain) and *Terminalia belerica* (Bahera) are planted in better soil conditions.

The Aravallis

Aravallis extend from Gurgaon and Faridabad districts to Mahendergarh and Bhiwani districts. The main species used for afforestation are *Acacia senegal* (Khairi), *Prosopis juliflora* (Kabli kikar), *Ziziphus species* (Ber) and *Acacia nilotica* (Kikar-Babul).

Saline and Alkali lands

In the case of saline alkali areas, mounds or ridges 30-60 cms are formed in advance and soil allowed to be leached in rainwater. Auger holes about 1 m deep are made if the soil is stiff or contain calcium carbonate pans. Species used are *Acacia nilotica* (Babul-Kikar), *Prosopis juliflora* (extremely difficult areas), *Pongamia pinnata, Terminalia arjuna* (Arjun), *Azadiracta indica* (Neem) and *Albezzia lebbek* (Siris).

Sandy Plain and Sand Dunes

Stabilized sand dunes are afforested by species like *Acacia nilotica* (Babul/ Kikar), *Ziziphus mauritiana, Azadiracta indica* (Neem), *Cassia siamea, Acacia tortilis, Albezzia lebbek, Dalbergia Sissoo, Pongamia Pinnata, Tecomella undulata* and *Eucalyptus* spp.

In active sand dunes, the main species planted is *Acacia tortilis. Erianthus munja* is planted in between for stabilizing the sand. Sowing of castor (*Decinus communis*) is also done in between the tree species.

Waterlogged Areas

Waterlogging for 3-4 months in a year during the rainy season is a common feature in parts of central Haryana, shallow areas can be rehabilitated by raising tree species on mounds or ridges. Depending on the requirements, 30 cm - 60 cm high ridges with a top width of 200 cm are made. In such areas, Eucalyptus, Arjun, Jamun (*Syzygium cumini*) and Wellows (*Salix* spp.) are used for planting.

A Pioneering State

Haryana has been one of the pioneering States to bring wastelands under tree crops. All the wastelands along roads, canals, railway lines and flood protection bunds have been taken up for raising trees. Panchayat areas, village common lands and institutional lands are being taken up for raising trees under Social Forestry Schemes. In addition to this, supply of seedlings to farmers at subsidized rates has been an important aspect of Social Forestry Schemes, with the result that farmers have started planting their wastelands (marginal agricultural lands and farm boundaries) with suitable tree species.

In the foothills of the Siwaliks, construction of small water harvesting dams for impounding the run-off and utilizing it for irrigation on a small-scale has improved the moisture regime of such eroded hill slopes and ensured the full cooperation of the villagers in the protection of the catchments. Villagers are capable of raising 2-3 crops with the water available, in place of a single rainfed crop in the past. It has also become possible to grow fruit plants and raise fodder crops in their small holdings, reducing their dependence on forests for their fodder.

A Few Suggestions

However, forests and trees cannot be protected in isolation. There is an imperative need for reducing the large number of unproductive cattle, and replacing them with more productive ones. The general economic condition of the rural poor needs improvement so that their dependence on small bits of unproductive agricultural lands and the forest lands is reduced. This is possible only by establishing small-scale industries in rural areas, provision of alternative fuel, popularizing fuel efficient hearths and conscious efforts to educate the villagers on environment.

7

Impact of Irrigation on Wasteland :

A Case Study of Gandak Command Area (U.P.)

S.S. Verma

A review of the role of agriculture in economic development on a conceptual plane starting from the "French Physiocratic" school of economic thought in the 18th century to the present-day thinking in this field as well as the historical course of development in many advanced countries leads to the conclusion that agriculture plays an important role in economic development as industry. Any effort at industrialization without concurrent modernization of the agricultural sector by making available the essential infrastructure gives rise to a dual economy as has been the case in India during the last three decades. However, in the developing countries like India agriculture produces the major portion of the national income.[1] Hence, the role of agricultural development in the total economy of the nation is undebatable. The present study aims at analysing the impact of irrigation which is an important infrastructure in the development of agriculture by increasing the net sown area with the maximum possible use of wastelands as arable land with reference to the Gandak Command Area (U.P.).

In addition to the physical factors, agriculture is also governed and largely influenced by the socio-economic condition of the region. The land tendency, size of holdings, social tradition, mechanization and equipment and infrastructural facilities influence the system of cultivation, the cropping pattern and cropping intensity which ultimately influence the agricultural productivity and rural development. But apart from the other socio-economic conditions, the availability of infrastructural facilities such as irrigation, transport, capital, market fertilizers and high-yield varieties of seeds impose varying limits of agricultural use of land, the crop grown and the agricultural production. But amongst all these elements irrigation is the most important ingredient for intensive and efficient cultivation and increasing the agricultural productivity through lessening the uncertainty and dependence on nature. Without proper water management for well-controlled irrigation round the year and efficient drainage facilities especially in the region where the period, amount and seasonal variation of rainfall are uncertain, uneven and unreliable, the use of fertilizers, high-yielding variety of seeds, insecticides and other new agricultural technology and techniques cannot provide the desired level of result.[2] The high yielding varieties give sufficiently good result even if fertilizer is not provided, proper irrigation is given to them. Thus, it is the irrigation which makes it beneficial to other agricultural inputs. Without irrigation a surfeit of other inputs would, therefore, be of no avail. Irrigation does not only boost up the degree of intensity and nature of diversification of agriculture but also increases the agricultural land by converting the agricultural wastelands into arable land for the production of crops by ascertaining the supply of water to irrigate the crops during the period of water deficiency. Thus, accepting the key role of irrigation in agriculture pursuits, the percentage of irrigated land to Net Sown Area is being taken an important index of agricultural development.

The vulnerability of India's agriculture owing to the vagaries of the monsoon has made it imperative to accord irrigation a high place in the national development plans. More than 600 major and medium projects were taken up after independence of which about 450 had been completed and other have started yielding partial benefits.

Study Area

The Gandak Command Area (U.P.) comprises all Padrauna and Hata tehsils of Deoria districts and Maharajganj tehsil of Gorakhpur district. Its total area is 6,732.37 sq. km^2 having a population of 4.06 million with a population density of 603 persons/km^2 which is much higher than the State's population density (377 persons/km^2). The area lies in the basin of the Great Gandak river. General topography is gently slopy and undulating. Many perrenial and seasonal rivers flow in the region in the north to south direction. It is a subhumid region with hot summer, wet pre-winter and moderate cool winter. The region has broad alluvial soil which covers the zonal differences, particularly in Khadar, Bangar and Bhat lands.

Agriculture is the mainstay of the region. About 81 per cent of the total area is under cultivation with a marked sub-regional variation. The northern and eastern parts of the region, which have relatively more land under forest, depressions, water bodies and marshes, record a lower percentage of the net sown area. Eighty-five and a half per cent of the total cropped is under food crops and 17.75 per cent is devoted to sugarcane which is the main cash crop of the region.

Gandak Command Area Development Project

The Gandak Command area has very huge population density (603 persons/km^2) resulting in small uneconomic holding and subjected to floods and droughts sometimes simultaneously. The Gandak Project, therefore, envisaged to convert the calamity into a blessing by controlling the floods which were a usual feature in the region and by providing the irrigation water to the fields. These are two major purposes of this project.[3]

The work of construction of the main Gandak Canal was started in the year 1960 after an agreement between India and Nepal which was signed in 1959.[4] The water of river Gandak is diverted into canal by constructing a 740 metre long barrage on the river at Bhaisalotan (now renamed as Valmiki Nagar) situated in Bihar. Its total capacity is 15,800 cusecs with U.P.'s share of 7,300 cusecs.

With the completion of construction of the main canal (112.5 km) and about 2,500 km distribution system, it was found necessary to ensure proper soil and water management to increase the agricultural production. To achieve this goal, the Gandak Command Area Development Project, a statutory in 1972 and the irrigation was started from the *Rabi* season of the same year. This made the project and the programme also more useful and effective. The Authority was given adequate powers and changed with the responsibilities of an all-round integrated rural development by carrying out multidimensional developmental activities including agricultural production, horticulture, fishery, sericulture, bee-keeping, animal husbandry programmes along with particular emphasis on fully utilizing the available irrigation potential.

Development of Irrigation

Till the first decade of the present century only 32.14 per cent of net sown area was irrigated which was concentrated in the south and south-west of the region.[5] Waterlogging had been a serious problem in the northern part of the region which resulted in low agriculture in the *Kharif* and late *Rabi* sowing, and the eastern part was a flood-affected region. Due to these natural calamities, the irrigation was not in vogue in those regions.

After independence there was increase in the population. Control on the floods and introduction of new agricultural techniques increased the importance of irrigation and gave a boost to the agricultural productivity to meet the demands of the growing population. However, in 1951 only 41.13 per cent of net sown area was irrigated by tubewells, wells and ponds. Most of the irrigation sources were operative in the private sector and were not adequate and available to the common people.

In 1972, the irrigation was started by the Gandak Project canals, and now there is a 112 km long main canal and about 3,029 km distribution system in the region which has an irrigation potential of 3.3 lakh ha.[6] It had also been experienced that existing irrigation potential would not be sufficient to cope with the water requirements of the command area. The project had been designed to irrigate only 75 per cent of net cropped area. The maximum potential is available in the *Kharif* season whereas the maximum demand of water is in *Rabi*. Hence, it is proposed to install 2,000 tubewells each of 3 cusecs capacity and 444 tubewells have been bored so far. Along with these, the financial incentive was made available to small and marginal farmers to install private tubewells and to purchase pump sets to create an irrigation potential of about 3.20 lakh ha by the end of the year 1983-84.

Changes in Irrigated Area

In 1970, the extent of the irrigated area was 2.18 lakh ha which was only 43 per cent of net sown area and by 1984 the total irrigated area was 3 lakh ha which is 58.3 per cent of net sown area. It is more than the State (52%) and national (25.8%) average.

There is a regional variation in the share of irrigated area to net sown area. In 1970, south-west, south and western parts of the region had maximum extent of irrigated area while central and north-eastern belt had least irrigation practice. During 1970-84, there has been a substantial growth in the irrigated area in every part of the region. However, the pattern of spatial distribution of percentage of irrigated area to net sown area in 1984 is more or less same as it was in 1970 because the north-eastern part still has very low extent of irrigated area due to lack of distributaries and minor gullies and the flooding river Gandak almost every year. South and south-western parts have very high percentage of irrigated land. The central part of the region which is known as Bhat has very high capacity of soil moisture storage. The stored moisture is utilized by *Rabi* crops during the period of water deficiency. Hence, the development of irrigation facilities not so necessary even for the *Rabi* crops.

The growth in the extent of irrigated area during 1970-84 is 41.96 per cent because in 1970 the extent of irrigated land remained 2.118 lakh ha which increased to 3.1 lakh ha in 1984. Thus, the absolute growth in the irrigated area was 22,000 ha with the positive growth of 41.96 per cent, while the growth in the irrigated area was only 28 per cent during the first seven decades of this century (1901-70) in the region. It reveals that after the commencement of the Gandak Project, an appreciable increase in the irrigated area was recorded.

The north-eastern part of the region which includes Khadda, Naurangia, Bisunpur, Sevrahi, Dudhahi Development Block recorded a higher growth in irrigated area because before the Gandak Project there was scarcity of irrigation facilities due to floods, waterlogging and Bhat soil but the Project provided the irrigational facilities which increased the extent of irrigated area. The western part experienced a medium growth rate while a low and lowest growth rate was recorded in the southern part of the region where irrigational facility was already developed, before the Gandak Project.

Source of Irrigation

The irrigated areas by different sources for both the years are given in Table 7.1.

Table 7.1 : Irrigated area by different sources

Source of irrigation	*Area in Ha*		*% of Irrigated area*		*Variation in (1970-84)*	
	1970	*1984*	*19701*	*984*	*Total*	*Per cent*
Canal	12262	168717	5.6	54.5	+156455	1275.9
Tubewell	99273	101142	45.5	32.7	+1869	1.88
Well	65912	16179	30.2	5.2	-49733	-75.45
Pond	28371	8753	13.2	2.8	-19618	-69.14
Others	12354	14918	5.5	4.8	+2564	-20.76
Total	**218172**	**309709**	**100.0**	**100.0**	**+91537**	

It is obvious from Table 7.1 that the canal irrigated area multiplied its extent 1,200 times during 1970-84. In 1970, only 5 per cent of the total irrigated area was irrigated by canal but now it is 54.5 per cent, while the irrigated area by tubewells has increased only by 1.88 per cent whereas the area irrigated by wells and ponds decreased to 75.45 per cent and 69.14 per cent. Thus, now the canal is the principal source of irrigation and its importance has increased in every part of the study region. But in the north-east and south-west part of the region, where canal irrigation is not available, tubewells are still the main source of irrigation.

Land use

The pattern of land use of any region is the cumulative effect of the physical and cultural factors and the type of ecological changes. The land use pattern of the study area is being shifted from subsistence agriculture to intensive subsistence agriculture which is apparent through the stage of land use and socio-economic organization of the region. The land use and its changes are recorded in Table 7.2.

Table 7.2 : General land use (1970 and 1984)

Land use	*Area in Ha*		*% Total area*		*Variation*
	1970	*1984*	*1970*	*1984*	*in %*
Arable land	507616	530420	73.3	80.75	+4.49
Agricultural wasteland	47545	22596	7.2	3.44	-52.2
Forest	52591	50382	8.0	7.67	-4.1
Not available for cultivation	49115	53469	7.5	8.14	+8.5
Total	**656867**	**656867**	**1000**	**100.00**	

It is obvious from Table 7.2 that the extent of arable land was 5.08 lakh ha in 1970 which was 7.3 per cent of the total area of the region whereas agricultural wasteland was only 47.5 thousand ha (7.2% of the total area). Forest and non-available for agriculture had 8 per cent and 7.5 per cent of the total area respectively.

The data in Table 7.2 reveal that in 1984 about 5.3 lakh ha (80.75% of the total area) were under arable land. The extent of arable land was increased by 4.4 per cent because 22,804 ha (3.47% of the total area) of additional land was put to crop production by the year 1984. Thus, the growth in extent of arable land is very slow because it had nearly reached the saturation point and no further growth could be expected.

The extent of the area not available for agriculture is 53,469 ha (8.14% of the total area) in 1984 which was 49,115 ha in 1970. Thus, it increased by 8.5 per cent during 1970-84. The extent of forest has decreased by 4.1 per cent and now its area is only 7.67 per cent of the total area.

Change in the Wasteland

In 1984, the extent of agricultural wasteland was 22,596 ha which is 3.44 per cent of the total area while it was 47,54 ha in 1970. Thus, it decreased by 52.2 per cent during the past 15 years because 24,949 ha of wasteland was put to other land uses.

The decreasing trend in the wastelands was experienced in every Development Block but six Blocks which are concentrated in the south-eastern part recorded higher decreases (more than 75%). This part was the flood-affected area of the little Gandak which is now controlled and most of the wasteland of the flood plains is now under crop production. Four Development Blocks of the eastern part experienced high decreases (60% to 75%) while medium decrease (45% to 60%) was recorded in seven Development Blocks of the western and north-eastern parts of the region, less and least decrease (<45%) was recorded in 16 Development Blocks of the western and central parts where the extent of wasteland was very small in 1970. Thus, a large tract of the eastern and north-eastern part along river Gandak which was lying uncultivated as wasteland due to frequent changes in the course of river, undulating high banks and the menace of regular flood and lack of irrigation facilities during the *Rabi* season has recorded a high decrease in the culturable wasteland because of a major portion of such area is now put to crop production by taming the river and by providing the irrigational facilities.

The Impact of Irrigation on Wasteland

As it has been analysed, a positive role is played by irrigation in the extension of agricultural land by utilizing more and more agricultural wastelands as well as in agricultural development. Now, there is a consensus that irrigation provides the base for utilizing the wasteland.

To find out the degree and direction of correspondence among the percentage of irrigated land and the wasteland, the correlation coefficient is calculated with the help of Kart Pearson's method, for which the percentage of irrigated land has been taken as independent variable (X variable) and percentage of wasteland as dependent variable (Y variable). The correlation coefficient between these two variables is — 5.26 for the year 1970 and 5.63 for the year 1984. Thus, both the variables have statistically significant (negative) correlation which establishes the hypothesis that the higher the extent and degree of irrigation, the lesser the extent of wasteland and *vice versa*. In this context, it may be inferred that irrigation is a prerequisite and a strong explanatory variable to minimize the extent of wasteland.

Sample Village Study

Six villages—Laukaria, Kishunpura, Piparia, Belva, Rampur and

Ahirauli—are selected for the empirical study of the impact of irrigation on wasteland and to prove the result drawn through the above general study. The villages are selected with various physical and cultural environment and they have different extent and source of irrigation. The general land use data of the villages for 1970 and 1984 are given in Table 7.3.

Table 7.3 : Land use of Selected Villages

Total Area	*Laukaria (392)*	*Kishunpura (687)*	*Piparia (221)*	*Belva (136)*	*Rampur (106)*	*Ahirauli (224)*
Net sown(1970)	363	554	204	117	48	190
Area(1984)	364	560	190	113	51	208
Waste-(1970)	11	76	2	8	4	22
land (1984)	8	69	00	2	2	00
Irrigated (1970)	00	144	143	100	17	95
land (1984)	364	179	190	113	26	28

The data indicate that except Rampur village, all the villages had more than 8 per cent of the total area under cultivation in 1970 and the extent of wasteland was very low. Maximum (11%) wasteland was in Kishunpura followed by Ahirauli (9.8%). Lowest extent of wasteland was recorded in Piparia (0.9%).

The data also reveal that there was a variation in the extent of irrigated land in 1970. Belva has the highest share of irrigated land (85%) followed by Piparia (70%) and Ahirauli (50%). There was a total absence of irrigated land in Laukaria in 1970.

A comparison of data of land use of 1970 and 1984 reveals that considerable changes were recorded in land use during 1970-84. The Net Sown Area of every village increased slightly except in Belva where it decreased by 3.4 per cent. A noteble change is experienced in the percentage of irrigated and wasteland. After the introduction of various means of irrigation, the extent and percentage of irrigated land increased by 13 per cent to 118.9 per cent in the selected village while a reverse trend is noted in the extent of wasteland because a considerable decrease is recorded during 1970-84. Now, there is total absence of wasteland in Piparia, Rampur and Ahirauli. The correlation between the percentage of irrigated land and wasteland is 5.3 and 7.2 for the year 1970 and 1984 respectively which reflects the same conclusion as it is inferred through general study.

The data reveal that a major portion of the wasteland of these villages is now being utilized as arable land which increases the extent of Net Sown Area of each village.

Conclusion

The above analysis of land use data of 1970 and 1984 of the whole region and selected villages can be summarized as that irrigation is the basic infrastructure for the growth of crop output by extending the arable land, increasing the cropping intensity, improving the cropping pattern and enhancing the land productivity specially in those areas where the amount and period of rainfall is uncertain, uneven, dispersed and unreliable. It is also inferred that irrigation facility is an important factor for converting the wasteland into arable land for crop production.

NOTES

1. Singh J. and Singh, B.R (1981). "Agricultural Intensity, Diversity and Rural Development : A Case Study of Gorakhpur Tehsil, *Uttar Bharat Bhoogol Patrika*, Vol. XVII, p. 9.
2. Verma, S.S. (1985). Urbanisation and Regional Development, Ph.D. Thesis (unpublished), University of Gorakhpur, p. 297.
3. Sen, Bandhudas (1974). *The Green Revolution in India : A Perspective*, Wiley Eastern, New Delhi, p. 16.
4. *India: A Reference Book* (1983). Government of India, p. 248.
5. Nevill, H.R. (1901). Gorakhpur: A Gazetteer, *The District Gazetteers of the United Provinces of Agra and Oudh*, Vol. XXXI, Allahabad, p. 39.
6. Gandak Command Area Development Authority, *Project Gorakhpur: At a Glance*, Nov. 1982, Information and Public Relation Wing of Gandak Project, p. l.

8

Structural Measures for Wasteland Development

A. C. Chaturvedi

The close relationship between water and vegetation was established early in France in 1215 during the reign of Louis VI. In 1342, a forest was protected against avalanche in Switzerland, and between 1535 and 1777 protection forests were proclaimed in 322 instances. Torrent control was taken up in 1860. The movement of an damage by shifting sand resulted in measures for stabilization as early as 1739. An Act providing for a penalty for grazing of the meadows near the beach was passed in Truro. In California, 75,00,000 hectares of land were reserved for watershed protection.

Erosion and Sedimentation

In higher altitudes the erosion causes soil particles to be detailed from the soil surface. This was found to be generally beneficial, because it proceeded at a slower rate than at which the new soil was being formed and it helped to keep the soil in a productive condition. Erosion, transportation and deposition were the typical features of the area. In lower reaches, sheet erosion was observed more or less uniformly over the entire soil surface. Plants were observed to be standing with part of the roof system exposed, indicating that the soil was washed away around it. Sheet erosion was observed to be responsible for damaging a greater land area. In large areas hill erosion was found to result from scouring of micro-channels in which water concentrated as it ran downstream. The grills merged to form gullies.

Economic Damages

The methodology for estimating the damage costs incurred by users of water resources is to be developed and application of the methodology has to be illustrated. The objective in preparing the methodology is to provide the principles and procedures required to assess the economic and social damages resulting from water use. Such damages occur in many different ways, some of which are easily identified and measured, whereas others are neither obvious nor quantifiable.

River and Drainage Control

The State irrigation departments are concerned with river and drainage control. Torrents in the upper catchments cause considerable damage. The hills in U.P. provide a classic example of effect of deforestation and erosion. The formulae for close relationship of vegetation and watersheet flow were established.

The Severity

The raindrops were observed to strike the ground at velocities of about 10 metres second and 2.5 cm of water over an area of ground was found to weigh more than 110 tonnes. These raindrop splashes were an important factor in erosion. This type of erosion was prevented and reduced by maintenance of a dense ground cover. Where the velocity of flowing water was doubled, its cutting power was increased four times, the quantity of material carried increased 32 times and the volume size of particles carried increased 64 times. Water fell down a 40 per cent slope with twice the velocity as that on a 10 per cent slope.

Modification of Vegetative Cover

Physical watershed characteristic and metrological processes were observed to affect streamflow varying by sections of the region and seasons of the year. Hence, the procedures for making the hydrologic evaluation varied according to these conditions, where the influencing factor of flood run-off was high intensity rainfall, the difference between floods that occurred without and with certain land treatment, cover condition changes were evaluated by observing the effects of the measures upon the infiltration and storages capacity of the soil. In snow melt regions, the cover effect was determined by an evaluation of the effect of the measures upon snow distribution, rates of snow melt and storage capacity of the soil.

Sedimentation

The contribution of sediment rate was found to be in inverse proportion to the catchment area. The absolute maximum rate for the major catchment was found to be within 4.33 hectare metre per 100 sq. km year. Some of the reservoirs have shown excessive rate of sedimentation, which was found to vary from 3.50 to 13.0 hectare metre/100 km year. It is recommended to evolve new silt index which would take into account the run-off. The siltation index of some of the small reservoirs in terms of old units was found to be lower in comparison to big reservoirs. The siltation index was fixed after capacity survey, and measurement of silt deposited. The siltation rate was estimated and the siltation index was established varying from 0.54 to 17.04.

Evaluation of Land Treatment Programmes

This was carried out under different hydrologic conditions:

1. The soils, vegetation use and condition of the watershed were surveyed to determine the area of all complexities of soil, cover and conditions of cover.
2. The area of the complexes in future, with and without a land treatment programme was estimated.
3. The stream flow from watersheds was estimated for vegetative conditions with and without land treatment programmes. The difference between the stream flow for the two conditions was a measure of the effect of land treatment measures.

The soil loss was found to be maximum in regions with high population density on account of pressure of food production. Over 50 per cent of the area was found to be suffering from erosion or other land defects like salinity and alkalinity, waterlogging floods and erosion. The land rendered unproductive due to development of alkalinity and salinity worked out to 3.0 million hectares. This is the largest area of its kind found in any single State in the country. These areas were located in the semi-arid regions of the State and the region lying south-west of river Ganga. This is being continually increased due to waterlogging caused by the newly constructed canals under new projects. A stretch of low-lying tract extends to the sub-humid parts of the State. These usar, Kallar or reh soils appear as extensive white, greyish white and coloured salt encrustation during dry periods.

Nature

These sterile and barren lands with hard rock-like surface are devoid of any vegetation, completely or in patches, and have poor or impeded drainages, with muddy water staggering over long periods during the rains. The surface was found to be impermeable in these soils, and a hard kankar layer was found at some depth. These salt-affected soils were quite variable in their occurrence. These small patches were also observed in fertile lands, where crop growth was stunted or completely missing, while at other places, they were found in belts of considerable expanse, interspersed with patches of fertile lands or continuously sterile. The alkali condition was found to change frequently at relatively short distances and within the soil profile. Plans were formulated for reclaiming about 10 per cent of the total area. A scheme of drains has been proposed for removing the harmful salts from the soil. The provision of dependable supplies of good quality water for leaching and crop manuring is a crucial input in development of salt-affected areas and removing the harmful effects from the soil.

Tubewells

Shallow tubewells are proposed for lowering the water-table and improving the sub-soil drainage. A high intensity of cropping is proposed in the areas. The canals were designed with high duty of irrigation. The underground water is planned for conjunctive exploitation for providing dependable supplies for reclamation. It is estimated that about 1.75 cm is the daily evaporation during off-season leaching period of May and June against 0.625 cm average percolation rate in the soil. The peak water requirements thus amounted to about 1.875 cm per day, one cusec of water was found to be sufficient for about 15 hectares of land. A shallow tubewell with a discharge of about a half cusec will cover an area of 7.5 hectares. The coverage through the tubewells will be double, where water from canals was also available.

Cost

The areas for reclamation were surveyed in detail for soil, hydrologic and topographic studies. After the areas and the problems were identified, the areas were bonded, levelled and ploughed 50 pc GR value. Gypsum was applied with water filled regularly every 10 days. The crops suggested were salt-tolerant paddy in *Kharif*, barley and wheat in *Rabi*. The average cost worked out to Rs.10,000 per hectare and Rs. 500 crores for 5 lakh hectares.

Flood Control and Drainage

The annual area affected by floods and waterlogging was worked out to 18.7 lakh hectares. An area of 7.5 lakh hectares has been protected so far. There is a plan for construction of embankments, new drainage channels, improvements to existing capacity of drains incorporated increasing capacity of rivers, extension of tight waterways, development of khadar areas with modernization and extension of flood forecasting services. The extension of waterways in bridges is also planned. Strengthening and raising of existing embankment is also planned. The cost per hectare, Rs. 2,500 and for protecting an area of 20 lakh hectares, Rs. 5,000 crores are needed. This is proposed to be covered in a period of 25 years. These works shall contain meandering of rivers, bank erosion and would help in checking the frequency and intensity of floods. Widest publicity of the schemes is suggested through the information media comprising the local press, radio and TV feature films are also proposed.

Environment

Protection of environment has come to constitute a sort of superior evolution or terms of reference of special laws, as those referring to water mining, groundwater pollution. Wherever new areas of irrigation are established, it does not naturally follow that water management has been carried out on scientific basis. In many areas, shallow groundwater was found to have high salinity due to evaporation and evapotranspiration. For saline waters, leaching of old salts on drain boundaries, deadened pores, small fissures, etc., have contributed to the salt load. Leaching is suggested for the salinization of groundwaters. Deep saline waters were found to have high bromine/chloride ratios, which were much higher than found in most shallow groundwaters in these and most other environments. Stable isotope data reflected low temperature equilibration with the host rocks.

Case Study

The Lalitpur area was investigated for salinity. The area under study was 500 sq.km and received 550 mm rainfall. The bedrock of the area was modified by erosion and deep weathering. Groundwater is recharged through outcrop around the catchment boundaries and was confined by tertiary deep weather profiles.

Ravines

Areas in the State having soil erosion problems leading to formation of ravines are extending over vast areas in the catchment of the Yamuna, Chambal, Betwa and their tributaries. The estimated forecasts indicate that 1,400 hectares are converted and added to the existing area every year which totals 12.3 lakh hectares. This is extending beyond the agricultural land to habitations, roads, railways and other public property. These ravines are providing escape routes to dacoits. Perspective plans were prepared so that land and soil resources were put to proper use, maximum returns were ensured and at the same time law and order problems were solved. The core programme of reclamation brought back the eroded area under reclamation and protected the tableland from conversion into ravine.

The medium and deep ravines were put under an afforestation and horticultural programme and shallow ravines were put under the plough. Necessary structural works of pitching of treated earth, pipe outfalls of local materials including timber were designed. Programmes were finalized to reclaim about one-third of the area of 4.23 lakh hectares. The average cost of reclamation including shallow, medium and deep ravines is Rs. 3,600 per hectare. Shallow tubewell in the upper reaches and deeper tubewells in the lower reaches are recommended for providing irrigation facilities. These works will cost Rs. 500 crores. The cost will go down considerably if work is entrusted to voluntary agencies.

Aiding the Poor Class Beneficiary

Far-reaching benefits will flow for the poorest section of society from the above programme. The landless shall be provided full employment in these schemes as the programmes envisaged are labour intensive. Sixty to seventy per cent of the expenditure will go in the form of wages. Additional employment is envisaged in collection, processing and manufacturing of the various products.

Training and Demonstration

For maximizing benefits, the programme is proposed to be carried out in each river watershed by locally organized village councils. The plans shall be framed in the villages by different action groups. Demonstrations of mendbunding, levelling, landshaping for conserving rainwater shall be regularly held by village team leaders, trained collectively for a group of 10-15 villages. The training will be regularly imparted for six months in a year, as the farmers will be allowed to remain busy for the remaining period with their personal and agricultural operations. These demonstrations shall convince the cultivators regarding the utility of the new technologies. Five demonstrations shall be held in each block.

Constraints

The development programmes in wastelands have constraints like

(1) leakages in lining of channels, (2) working of shallow tubewells, (3) breaches in bunds, (4) silting of drains, (5) subsidence of pitching and (6) delay in execution of schemes.

Involvement of the local people shall ensure removal and quick repair of rural facilities. Small pamphlets of a few pages shall describe these techniques in an easily understandable language. The beneficiaries will operate the water delivery arrangements. They would in this case have the best utilization of hydraulic resources by limited and judicious application and distribution of water, thus resulting in increased crop yield. The experience of State run schemes has repreatedly demonstrated defects in channel lining, embankments, working of shallow wells even during construction. These works were found to cause havoc when put in operation. This has often belied the expectation of the beneficiaries who are generally becoming suspicious of government schemes.

Regulation

The wasteland development works executed by the beneficiaries under the guidance of both the State and voluntary agencies will have adequate operational standards and emergency preparedness carefully planned and implemented.

Increased Efficiency of Shallow Tubewells

The survey carried out by the author indicated operational efficiency of tubewells varying between 15 and 25 per cent, one-third of the tubewells were found having efficiency around 40 per cent, another one-third had efficiency of 40 to 8^ per cent and the remaining one-third 7 to 20 per cent.

The reasons for low efficiency as analysed were:

1. Excessive suction head,
2. Use of sharp bends in the piping systems,
3. Improper selection and inferior quality of pump,
4. Excessive height of delivery pipe from ground level,
5. Improper fixation of joint,

6. Carelessness in annual overhauling of the pumping set,
7. Leakage in joints,
8. Lowering of an undersized pipe in the suction line, in case of leakage in the original suction pipe,
9. Non-matching of the revolutions of the pump and motor,
10. Use of poor quality of the driving belts,
11. Improper size of suction and delivery pipe,
12. Poor quality reflex value, and
13. Unlevelled plinths and foundations and misalignment of pump and motor pulley.

The efficiency of pump sets was improved to 50 per cent in 15 selected Blocks of Central and Western U.P. The cost involved varied from Rs. 30 to Rs. 14. There was additional energy saving of 15 per cent. The masonry pit was dig deeper to the desired level and resulting energy saving was 5 per cent. The sharp bends in the delivery pipes were replaced. The investment involved for one tubewell was Rs. 60 only. High end suction pumps replaced ordinary pumps. The suction head ranged from 5 to 8.3 metres. This factor alone increased the operational efficiency by over 30 per cent. Rigid PVC pipes were provided to replace G.I. pipe both in suction and delivery. The electricity demand was curtailed by 65 per cent. A properly selected reflex value replaced the generally installed poor quality reflex values. These techniques are recommended to be adopted on a pilot basis to increase the efficiency of existing works.

Involovement of Masses and Voluntary Agencies

The involvement of masses and voluntary agencies is needed: (1) To transfer research and demonstration techniques to the field, (2) Suitable modifications in social and forestry programmes through output sharing and ownership concepts should be made, (3) Results of relevant scientific studies for building up a strong information base which will help in proper management of resources be made available, (4) Technology for construction of new programmes of road building, widening and corner cutting activities be made available and (5) Schemes for rehabilitation of 200 million people on either side of the road be drawn up.

Role of Models

The problem of wasteland development and role of models in the management of water quality are important. The mathematical water quality model comprised three sub-models, flow, mass transport and heat transport. The formulation of sub-model was based on the continuum approach. The relevant microscopic equations wave in terms of measurable continuous quantities. The aquifers dispersivity was shown to be an increasing function of the scale of averaging. The formulation of the model was connected to the boundary conditions and groundwater quality forecasting problem and helped in augmenting the qualitative and quantitative results.

Recommendations

It is recommended that base-line survey for all development projects and specially irrigation projects may be taken up one year prior to the start of the project as to cover the whole year. The evaluation is recommended. This may be followed by catchment area and watershed management. Research and development is recommended on physical and biological aspects of rehabilitation together with regulation and management of water. The performance of existing works in the private sector is proposed to be stepped up by about 50 per cent in five selected Blocks in each district. All districts of the State are proposed to be covered in a period of 5-10 years.

REFERENCES

Chaturvedi, A.C. (1975). "Pilot Survey for Area Planning and Regional Development", *Indian Journal for Power and River Valley Development,* Calcutta, p. 12.

Chaturvedi, A.C. (1976). "Structural Designs and Problems of Water Pollution Control," *Proceedings,* International Conference, Asian Institute of Technology, Bangkok, p. 408.

Chaturvedi, A.C. (1977). "Water Quality and Availability in India", *Proceedings,* Third International Hydrology Symposium, Agricultural Research Centre, Fort Collins, USA, p. 128.

Chaturvedi, A.C. (1978). "Storm drainage in India", *Proceedings* International Conference on Irrigation in Wetlands, Agricultural University, Wagenin, Netherlands, p. 48.

Chaturvedi, A.C. (1981). "Irrigation in Indian Cultivation", *Proceedings,* International Conference on Irrigation in Wetlands, Agricultural University, Wagenin, Netherlands, p. 48.

Chaturvedi, A.C. (1979). "Land Erosion and Sedimentation", *Proceedings,* International Symposium on Erosion and Sediment Transport Measurements, Florence, Italy, p.18.

Chaturvedi, A.C. (1985). *Controlling Reservoir Siltation in 1980s*, International Conference, Preserve the Land Soil Conservation Society of America, College of Tropical and Human Resources, Agricultural University, Hawaii, p. 31.

9

Evaluation of NRSA Wasteland Map from User's Angle : A Case Study in Meerut, Ghaziabad and Bijnor District of Uttar Pradesh

D.B. Misra, S.L. Dabral and *A.K. Gulati*

Land is, perhaps, the most important resource of a country. When used scientifically, the country enjoys perpetual prosperity. If we look at the culturable land resource position of the world, we will find that India is at the top of it. However, a great part of the land in the country, especially on the Deccan plataeu and central high land, has been rendered unproductive due to unscientific use over ages. The new 20-point Programme announced by the Prime Minister envisages a new strategy for forestry. There is a target to afforest five million hectares of land annually. The bulk of the area for afforestation is expected to come from what at present is generally known as wasteland.

Afforestation of the wastelands is expected to restore the ecological balance, improve the environment and make the village self-sufficient in fuel and fodder production. Self-sufficiency of villages in fuel and fodder will free our forests from enormous destructive pressure.

Different agencies have estimated the extent of wasteland from time to time in varying manner with different objectives. Some of these agencies are the Ministry of Agriculture, Directorate of Economics and Statistics, National Wastelands Development Board, etc. The latest effort made in this regard is by the National Remote Sensing Agency (NRSA), which was commissioned by the Government of India for estimating the extent of wasteland in the country with a view to making available basic data to the National Wastelands Development Board. The effort of the NRSA evoked wide interest as it used satellite remote sensing technique for this purpose for the first time in India.

The NRSA prepared wastelands maps of different states on 1:1 million scale and of the whole country on 1:3.5 million scale by visual interpretation of Landsat imageries for the period 1980-82.

Definition

Wasteland as defined by the NRSA is that land which is presently lying unused or which is not being used to its optimum potential due to some constraints. The definition suffers mainly from two drawbacks. One, the fallow land which actually is not wasteland will be classified as wasteland. Two, almost all lands can be classified as wasteland, since the lands are rarely used to their optimum potential. The more acceptable definition has been adopted by the National Wastelands Development Board according to which the land will be considered as wasteland, if it is degraded and is presently lying unutilized due to different constraints.

According to the nature of constraints, the wastelands are classified into two broad classes: culturable and non-culturable.

The culturable wastelands are those which at present are lying unutilised but are capable of being brought under such uses as agriculture, pasture and forests. On the other hand, the non-culturable wastelands are all those which are at present barren and cannot be put to any productive use. Examples of such wastelands are: rocky outcrops, snow-bound areas or glacial areas.

Description of Study Areas and Objectives

The present study was undertaken to determine (i) the interpretation accuracy of the wasteland maps prepared by the NRSA in Bijnor, Meerut and Ghaziabad district of U.P. and (ii) suitability of 1:1 million scale for mapping the wastelands.

The geographical area of the districts falling under the study are is approximately 11,418 km^2. The terrain is mostly plain with slightly undulating upland occurring in south-south-west of Bijnor district. In these districts, the wasteland mainly falls in the category of salt-affected, sandy and waterlogged areas. The soil is mainly deep alluvial. Patches of saline soil occur in poorly drained areas. Sandy soils are found near the rivers and in south-eastern parts of Bijnor district. The Ganga with its tributaries, the Ram Ganga and Kho, flow through the tract. The Yamuna forms the boundary of Meerut district with Haryana. These districts are covered by a network of the Ganga Canal system. A belt of land known as "Khader" occurs all along the Ganga and its width extends to a few kilometres on both sides of the river. This "Khader" belt contains typical sandy and waterlogged areas. Soils in the area are mostly neutral with pH ranging from 6.5-7.5. The Khader belt is slightly alkaline soils with pH value above 8.5 generally do not exist. The land use pattern, according to district statistics (1980-81), is given in Table 9.1.

Methodology

The NRSA map pertaining to the study area was checked on the ground during August this year to find out:

1. Whether the patches shown as wasteland really contained any patch of the wasteland exceeding 1 km. in width.
2. If so, whether it was classified under correct category of the wasteland.
3. In the areas shown as non-wasteland, whether wastelands exceeding 1 km. width existed or not.

In the case of the 56 wasteland patches on the NRSA map, 100 per cent checking was done. Non-wasteland area was checked by the random sampling method. For this purpose, the map of the study area was divided into 2 mm × 2 mm squares. The squares were numbered for the purpose of drawing random samples of non-wasteland area. In the non-wasteland stratum, random samples of 2 mm × 2 mm (2 km × 2 km on the ground) equal to the number of wasteland patches, i.e. 56 were selected for ground truthing. The patches so selected were located on the ground by enlarging the 1:1 million scale map to 1:2,50,000 scale and using the corresponding topographic sheets on the same scale.

It was not possible to comb every hectare of the wasteland patches. In order to verify whether the wasteland patch shown on the map existed

Table 9.1 : Land use pattern

District	*Total Geog. area (ha)*	*Forests (ha)*	*Agriculture (ha)*	*Land under Misc. tree crops & groves*	*Other uses (ha)*	*Pasture (ha)*	*Waste-lands* (ha)*
Bijnor	491591	70847	353338	3587	50670	701	12448**
Meerut	391599	7992	325968	443	43950	610	12636
Ghaziabad**	–	–	–	–	–	–	–

* Village community lands have been included in wasteland.

** Figures for Ghaziabad district (part) which earlier formed part of undivided Meerut district were not available.

on the ground or not, the patch as per map was traversed across one of its long diameters and observed up to about 1/2 km on both sides of the line of traverse. The same procedure was followed in the case of non-wasteland patches. The scattered patches less than 1 km in width were considered part of the major surrounding type of the classification. This was necessary as the minimum mapping unit was of 1 mm width, i.e. 1 km/on the ground. Wherever more than 1 km wide wasteland/non-wasteland patches were found, the position shown on the map was accepted as correct. Estimate of the extent of the wasteland patch on the ground was not attempted due to time constraint.

Estimation of Interpretation Accuracy of the Maps

The observations have been summarized in Table 9.2.

Table 9.2 : Interpretation accuracy of maps

Category	*No. of patches on the map*	*No. of patches found on ground*			
		Wasteland			*Non-wasteland**
		Salt-affected	*Sandy areas*	*Waterlogged*	
(a) *Wastelands*					
Salt-affected	45	26	–	–	19
Sandy areas	11	–	5	3	3
Waterlogged	–	–	–	–	–
(b) *Non-wasteland*	56	6	1	2	47
Total	**112**	**32**	**6**	**5**	**69**

* Patches found in non-wasteland category were in agricultural use.

Table 9.3 gives the errors of omission and commission of interpretation for various categories of lands as a percentage of total number of patches shown on the map and found on the ground.

Table 9.4 gives the contribution of various categories of wasteland to the total errors of omission and commission.

Table 9.3 : Commission and omission errors of interpretation

Category of land	*Commission error (%)*	*Omission error (%)*
Salt-affected	42	19
Sandy	55	17
Waterlogged	nil	100
Non-wasteland	16	32

Table 9.4 : Contribution of categories of wastelands to the total commission and omission errors

Category of land	*Commission Error*		*Omission Error*	
	Frequency of misinterpretation	*% of total error*	*Frequency of misinterpretation*	*% of total error*
Salt-affected	19	76	6	67
Sandy	6	24	1	11
Waterlogged	–	–	2	22
Total	**25**	**100**	**9**	**100**

Table 9.5 gives the interpretation accuracy of the wastelands (category-wise) and is compared with the interpretation accuracy estimated by the NRSA.

Table 9.5 : Interpretation accuracy of the wastelands

	Present study		*NRSA report*	
Category	*Interpretation accuracy as %*	*Standard error*	*Interpretation accuracy as %*	*Standard error as %*
Wastelands				
Salt-affected	58	–	92.50	± 6
Sandy areas	46	–	88.24	–
Waterlogged areas	–	–	95.38	± 6
Non-wastelands	84	± 16	100.00	–

1. In the case of wasteland categories standard error has not been calculated as the accuracy of interpretation is based on total population.
2. The same formulae as used by NRSA for estimating the standard error have been adopted for calculating standard error of accuracy estimate of non-wasteland.

Results and Discussions

The National Atlas and Thematic Mapping Organization (NATMO) has compiled the Atlas of Agricultural Resources of India. In the soil type map of the study area, the land affected by salts has been shown. The patches shown as salt-affected in the Atlas are distributed almost on the same pattern as the salt-affected wasteland category shown in the NRSA map. The soil Atlas of NATMO is based on a chemical analysis of the soils. Such soils are not necessarily wastelands. This is found to be so during the ground visit of the study area as is evident from Table 9.2. It seems that the judgement of the interpreter has been influenced by the soil map of the NATMO.

Remote sensing based in Satellite imageries is still a developing science. In spite of this, a high degree of accuracy can be achieved by those interpreters who are well conversant with the ground truth and know about the seasonal variation in the ground cover. The facts revealed in the present study amply illustrate this point. The soils in the study area are fertile and valuable and in spite of slight salinity/alkalinity here and there, they are being successfully used to raise high yielding agricultural crops. If an agricultural scientist were to carry out an interpretation he would never show these saline/alkaline patches as wasteland.

It had not been possible to find out the season of the imageries used by the NRSA. It seems that the white/grey tone of the land due to post-harvest season or land being freshly ploughed or remaining uncultivated during summers, etc., has been mistaken for classifying the same as salt-affected areas. The errors of commission being as high as 42 per cent, illustrate this point. The percentage in the case of sandy areas is still higher, 55 per cent. The errors of omission for these classifications are, however, much less. The waterlogged areas, on the other hand have somehow escaped detection and have been misclassified as sandy areas/non-wasteland.

Table 9.4 reveals that the most misinterpreted areas (errors of commission) are non-wastelands, i.e. the agricultural fields have been shown as salt-affected areas (76%). The misinterpreted sandy areas have contributed 24 per cent to the errors of commission. Half of such areas on the ground were found to be agricultural land and the other half as waterlogged areas. Salt-affected areas contributed the maximum (67%) to the frequency of overall errors of omission followed by waterlogged (22%) and sandy areas (11%). The total contribution of the above categories of wastelands to the total error of omission as a percentage of the total number of patches is, however, small being only 8 per cent as compared to the total errors of commission (22%).

Interpretation accuracy in the present study has been found to be much less reported by the NRSA for entire U.P. for these classifications. This is evident from Table 9.5. As against 88.24 per cent and 92.50 per cent accuracies claimed for sandy and salt-affected areas respectively, the corresponding accuracies were found to be 46 per cent and 58 per cent.

Suitability of 1:1 Million Scale Maps for Estimating Wastelands

The study area lies in the Indo-Gangetic plain. Here the hunger for agricultural land has brought almost all the available culturable land under the plough. Rarely a patch of wasteland exceeding 1 km in any direction can be found in this region. Under these circumstances patches less than 1 km in width will get generalized with surrounding predominant agricultural areas on a 1:1 million scale map. As a matter of fact, most of the wasteland patches will be lost in generalization and will appear to be agricultural land. Thus, the map on 1:1 million scale will exclude majority of the wasteland areas existing on the ground and will depict a situation far from reality.

The NRSA has suggested another possibility, that a patch showing wasteland may consist of number of small patches less than 1mm × 1mm squares of other land within the same patch, resulting in significant generalization factor with the same, i.e. the patch appearing as wasteland on the imagery. However, during this study we did not come across such a situation. Instead, the situation is the other way round in the field, i.e. wasteland patches merge with agricultural field. Considering the above facts, it can be safely concluded that the wasteland map on 1:1 million scale will hardly be of any value to the users or the planners. Therefore, to get an acceptable estimate of the extent and reliable location of the wastelands, the minimum mapping unit should be comparable with the size of the wastelands occurring most frequently on the ground. A minimum mapping unit of 0.25 ha is expected to meet this requirement. This corresponds to a mapping scale of 1:50,000 and should be adopted for wasteland mapping.

It will not be out of place to state that the MSS data is not very suitable for mapping wasteland. Instead TM or some better data of carefully selected season should be used. It is also desirable that the project of mapping wastelands should be executed by the organizations connected with such surveys in order to ensure correct depiction of the facts.

Conclusion

(1) The interpretation accuracy of the NRSA wasteland map has been found to be considerably low because of errors of omission and commission.

(2) The 1:1 million scale is unsuitable for wasteland mapping. Mapping on 1:50,000 scale depicting wasteland patches down to 0.25 ha in extent is suggested.

(3) Better resolution data like TM data of a carefully selected season should be used.

(4) The wasteland mapping should be done by the organizations connected with such surveys to avoid anomalous interpretations.

Acknowledgement

The authors acknowledge with thanks the support extended by the Divisional Forest Officers of Bijnor and Meerut during the field visits.

10

Identification and Mapping of Wastelands at Micro-Level Using Remote Sensing Technique : An Approach

C.B.S. Dutt, P.P. Nageswara Rao, B. Manikiam, A.K. Gupta, G. Behera and *K Ganesha Raj*

The mainstay of India's economy land occupation of most of its population is agriculture. The rate of food grain production is not in proportion to population growth. The per capita cultivable land in India is only 0.26 ha (NWDB,1986). Land degradation both in cultivable and uncultivable lands due to deforestation, overgrazing, unscientific agricultural practices as well as natural calamities lead to the formation of wastelands. A wasteland is defined as "that land which is degraded and is presently lying unutilized except as current fallows due to different constraints". (Ref: NWDB,1986). Rapid growth of population, desire for better quality of life and the need for accelerating economic growth of our country necessitates the exploitation and reclamation of all wastelands which are at present either lying unused or which have been turned unproductive. Hence, the immediate task is to have timely and accurate information regarding the location and extent of all such lands known as "Wastelands". These lands may be under State occupation, private occupation or notified forest areas. Several studies on wasteland in India have been carried out by the Ministry of Food and Agriculture (1961), Planning Commission, Government of India (1963 and 1965), Ministry of Home Affairs (1972), and others. However, there is no accurate estimate of wastelands available mainly because of the differences in definitions and nomenclature of wastelands. There has been no attempt made to prepare a consolidated map on any scale showing the geographical location and spatial distribution of different types of wastelands occurring in the country. Organizations like the All India Soils and Land Use Survey (AIS & LUS) and National Bureau of Soil Survey and Land Use Planning (NBSS & LUP) have been carrying out soil mapping on different scales showing the occurrence of saline/alkaline, waterlogged and eroded lands in different parts of the country. As per the reports of the Directorate of Economics and Statistics, the area under barren and unculturable land is around 202.7 lakh ha. However, the above statistics does not give the totality of wastelands in the country.

The varying statistics is evident from Table 10.1 wherein the estimates of wastelands compiled by different agencies are shown. These estimates represent only the statistics but fail to give their spatial distribution. For the first time, the National Remote Sensing Agency (NRSA) of Department of Space, Government of India, prepared a nationwide wasteland map with an eightfold classification. The NRSA wasteland classification is based on visual interpretation of satellite data. The NRSA wasteland map and the estimates given therein formed the primary data base for identifying the districts with high incidence of wasteland. The Government of India has constituted a National Wasteland Development Board with the objective of bringing wastelands of the country under productive use through a massive programme of afforestation and tree planting and to monitor various wasteland developmental programmes. A Technical Task Group was constituted by the Planning Commission, Government of India for standardization of the definition and classification of wastelands. Wasteland categories and detailed explanation of each category as defined by the Technical Task Group are given in Annexure 10.1.

Having achieved a clarity on definition and categorization of wastelands, the next step is to develop a methodology to identify wastelands. at the micro level (forest areas, revenue lands and degraded farm land) along with the ownership details which is one of the end objectives desired by the National Wasteland Development Board. This comprehensive database will serve as a benchmark for future policy measures. The immediate priority is to prepare an inventory of wastelands for 182 districts where wasteland concentration is high. The present study is an attempt to evaluate the use of high spatial resolution remotely sensed data from spaceborne platforms for identification and mapping of wastelands at micro-level by integrating the Landsat Thematic Mapper data with revenue records.

Table 10.1 : Statistics on wastelands

Sl. No.	*Agency/Authors*	*Year*	*Reported Wasteland Area (million hect.)*
1.	Ministry of Agriculture	1979-80	17.5
2.	Directorate of Economics and Statistics	1978-79	38.5
3.	B. B. Vohra	1980	200
4.	Bhumbla and Khare	1984-85	93.7 (only non-forest area)
5.	National Land Use and Wasteland Development Board	1985	123

Methodology

The study area comprises an area of about 2,000 sq. km covering parts of Bangalore district (Karnataka State). Landsat Thematic Mapper Image (FCC) of March 15, 1985, was used for visual interpretation. The date of imagery is such that during this period most of the non-wasteland categories like reserved forest, agricultural plantations are at their maximum greenness stage while the dryland crops are already harvested thereby the spectral confusion with the wasteland categories is minimum. Bulk corrected false colour composite on 1:62,500 scale (approximately) has been used for delineation of wastelands. All the wasteland categories have been identified as per the classification developed by the National Wasteland Development Board, India (1986). After interpreting the wasteland details on satellite imagery, the same have been transferred in a base map of 1:50,000 scale using an Optical Reflecting Projector (ORP). The samples were randomly selected in the study area covering all the wasteland categories for estimating the interpretation accuracy.

The village boundaries as shown in Taluk maps available on 1:63,360 were transferred on to the wasteland map on scale 1:50,000. After transferring the village boundaries on the wasteland map, it was possible to identify the villages having the wastelands.

As per the standard random sampling technique (Cochran, 1977) at 95 per cent confidence level and 95 per cent accuracy, a sample size of 114 was arrived at. However, considering the accessibility, logistic and time constraint a sample size of 99 was decided upon for ground checking. Thirty-eight villages were randomly selected to check the locational accuracy of the wasteland identified at village and quadrant level (NE, NW, SE and SW portions of the village). The wasteland details from the 1:50,000 scale map were transferred to cadastral map (scale 1:8,000) for all the selected villages using ORP and the ownership details of the wastelands were checked with the help of information available with the village Karnams. The block diagram of the methodology developed and adopted in this study is given in Figure 10.1.

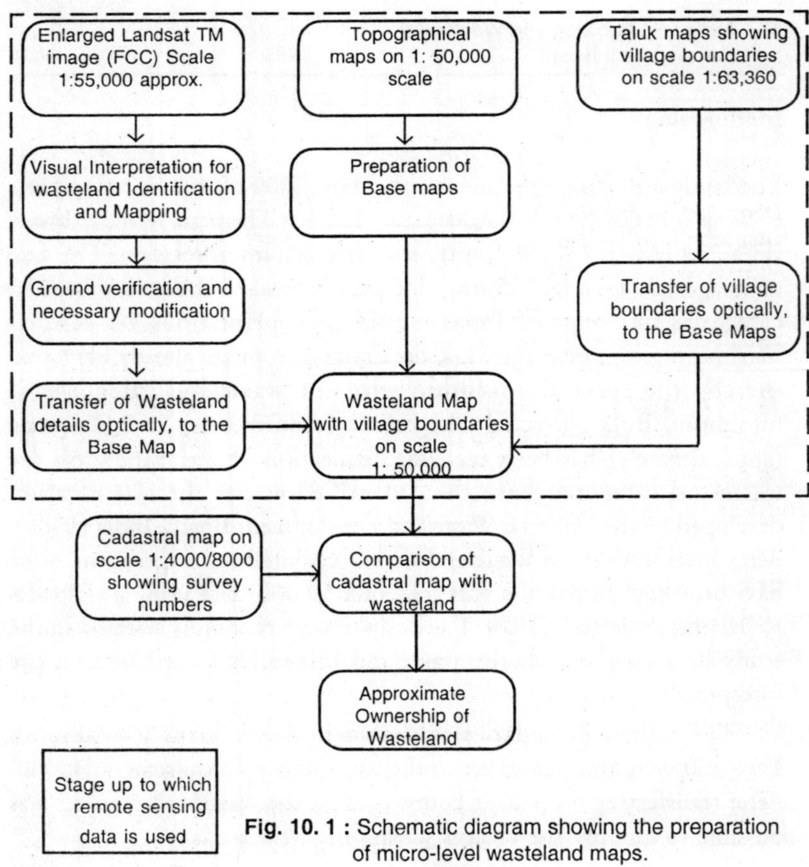

Fig. 10. 1 : Schematic diagram showing the preparation of micro-level wasteland maps.

Results and Discussions

The aerial estimates of wasteland category were measured on 1:50,000 scale wasteland map prepared for the study area using millimeter dot grid and are given in Table 10.2. It can be seen that the area under wastelands (about 154 sq. km) occupies 7.7 per cent of the total study area. A part of the wasteland map with village boundaries overlaid is shown in Figure 10.2. Data for category-wise interpreted vs. observed wasteland is given in the form of error matrix. (Table 10.3). The overall interpretation accuracy was found to be 76.7 per cent.

Table 10.2 : Wasteland Area estimates based on Visual Interpretation of Thematic Mapper FCC Data on 1:50,000 scale

Sl. No.	*Wasteland category (No.)*	*Area in sq. km*
	Culturable Wastelands	
1.	Gullied and/or ravinous land (1)	12
2.	Undulating upland with or without scrub (2)	55
3.	Surface waterlogged land and marsh (3)	15
4.	Degraded forest land (6)	6
5.	Degraded pastures/grazing land (7)	7
	Unculturable Wasteland	
6.	Barren rocky/stony wastes/sheet rock area (12)	59
	Total	**154**

The locational accuracies of wasteland categories at gross village level and at the village quadrant level have been assessed. It was found that out of 38 villages ground checked, 37 villages were found to have wastelands, thereby giving village level locational accuracy of 97 per cent. Similarly, out of the 68 quadrants verified on ground, occurrence of wasteland was confirmed in 67 quadrants giving an accuracy of 88 per cent. From the above results, it is clear that the Landsat-TM data can be used to generate wasteland data base at village level with an accuracy of 85-90 per cent. The cadastral maps of four villages with identified wastelands are given in Figures 10.3(a) to 10.3(d).

The interpretation accuracy of each category is also shown in Table 10.3. The misinterpretation of waterlogged areas was mainly due to the use of a single data TM-data when the waterlogged areas are filled with

Fig. 10.2 : Wasteland map of the parts of Bangalore district using landsat TM data

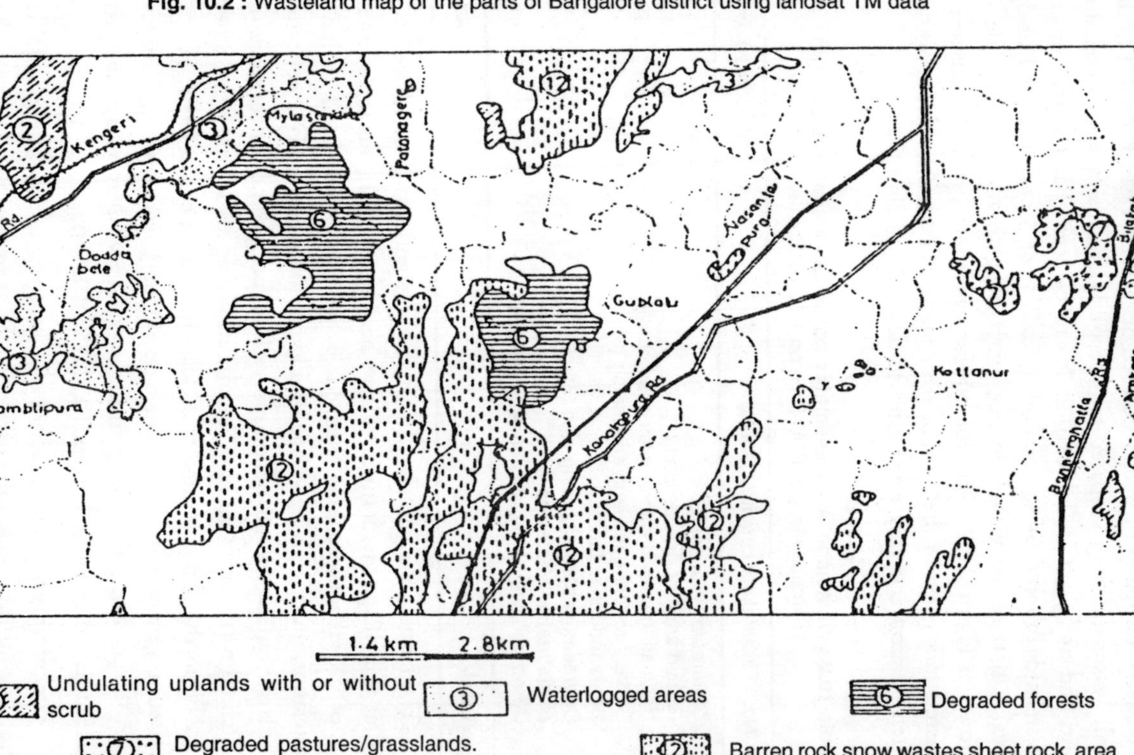

Table 10.3 : Error Matrix of Wasteland Categories Observed Wasteland Category

Category of W.L.	*1*	*2*	*3*	*6*	*7*	*8*	*12*	*Mis*	*Row Interp.*	*Interpretation accuracy (%)*
1.	14	–	–	–	–	–	1	1	16	87
2.	–	15	–	–	–	–	4	–	19	78
3.	–	1	5	–	–	–	–	8	14	35
6.	–	–	–	5	–	–	1	–	6	83
7.	–	–	–	–	7	–	–	2	9	77
12.	1	2	–	–	1	1	30	–	35	85
Column Total	15	18	5	5	8	1	36	11	99	76.7
Interpretation accuracy	93	83	100	100	87	100	83	–	–	–

Wasteland Categories 1 = Gullies; 2 = Undulating uplands with or without scrub;
3 = waterlogged; 6=Degraded forests; 7=Degraded pastures/grasslands
8 = Steep slopes; 12=Barren hills/sheet rock.

Fig. 10.3 (a) : W.L. Wasteland

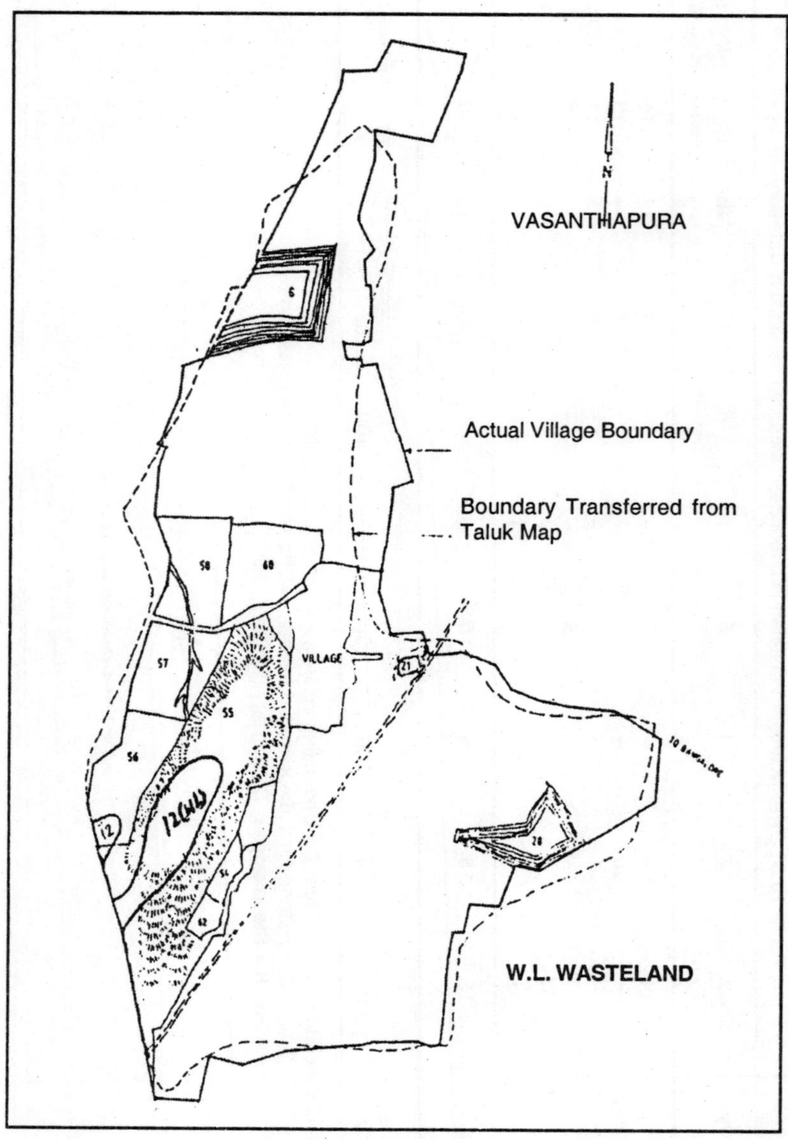

Fig. 10.3 (b) : Arakere

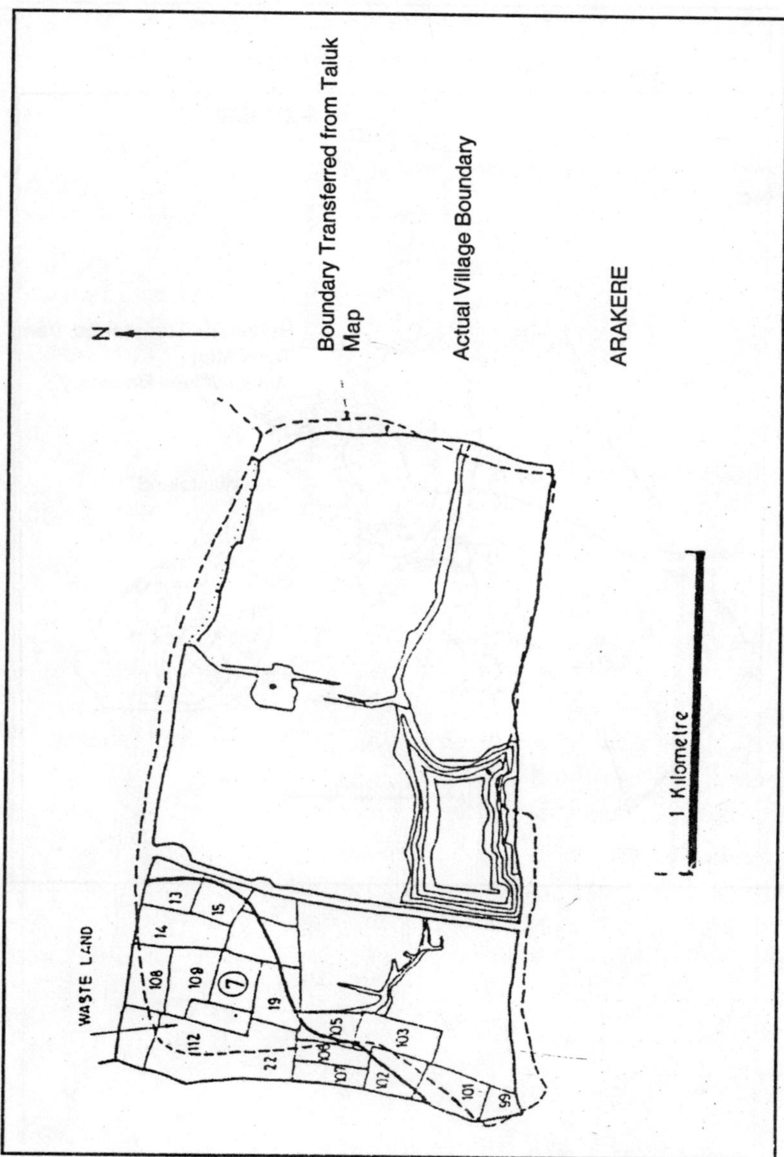

Fig. 10.3 (c) : Kengeri

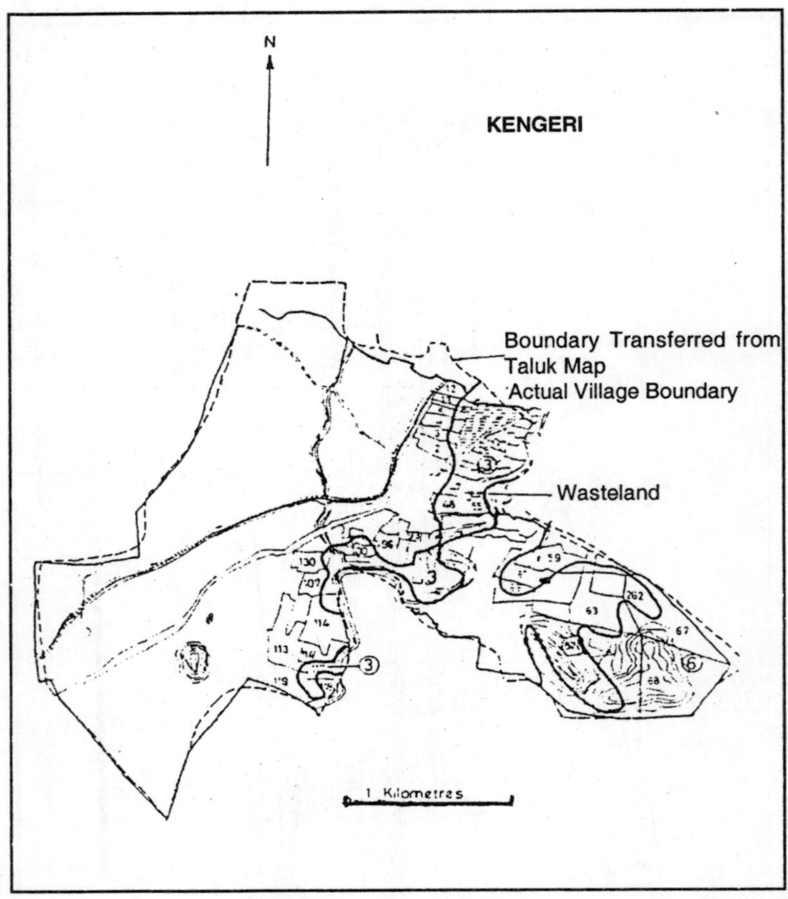

Fig. 10.3 (d) : Bellakana Halli

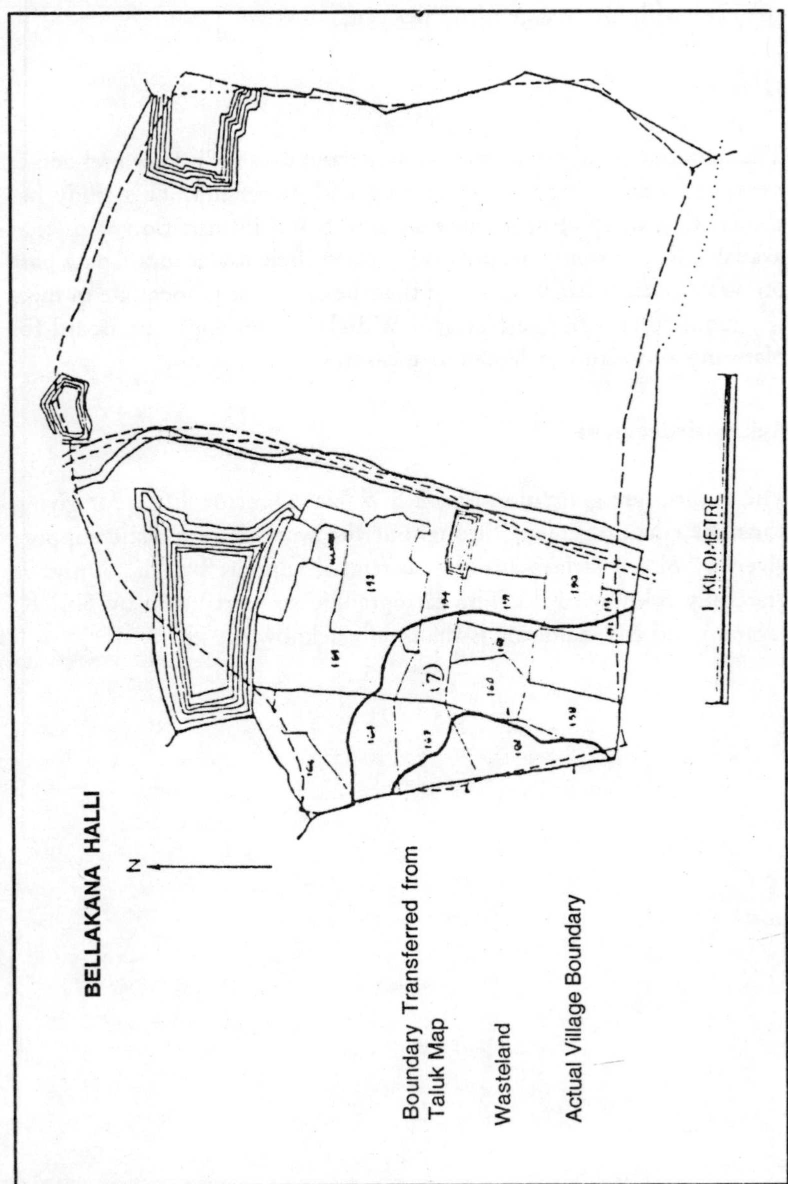

water hyacinth (spectrally similar to paddy). Excluding the waterlogged category, the interpretation accuracy was found to be between 77-87 per cent with an average of 83 per cent.

Conclusion

The developmental programme of wasteland calls for micro-level details on wastelands giving extent, location and ownership. This study has shown that integrating remote sensing based information with that available from revenue records can yield sufficiently accurate data base on wastelands. It has been found that the data base is adequate to meet the requirements of the National Wasteland Development Board for planning wasteland reclamation measures.

Acknowledgement

The authors are grateful to Shri Y.S. Rajan, Director, EOS, for giving constant encouragement throughout the work. The logistic support given by Shri J.S. Jayaram and secretarial support by Ms. Suniya is gratefully acknowledged. The cartographic support given by Shri K. Jayaram and Ms. Sameena is thankfully acknowledged.

ANNEXURE :10.1

Wasteland Categories Defined by the Technical Task Group

Level I		*Level II*
Culturable Wasteland	1.	Gullied and/or ravinous land
	2.	Undulating upland with or without scrub
	3.	Surface waterlogged land and marsh
	4.	Salt-affected land
	5.	Shifting cultivation area
	6.	Degraded forest land
	7.	Degraded pastures/grazing land
	8.	Degraded non-forest plantation land
	9.	Strip lands
	10.	Sands
	11.	Mining/industrial wasteland
Non-culturable wasteland	12.	Barren rocky/stony wastes/ sheet rock area
	13.	Steep sloping area
	14.	Snow covered and/or glacial area.

Explanation

Culturable Wastelands

Land which is capable or has the potential for the development of vegetative cover and is not being used due to different constraints of varying degrees is termed as culturable wasteland. Culturable wasteland comprise the following categories.

Gullied and/or Ravinous Land

Gullied Land : The gullies are formed as a result of localized surface run-off affecting the friable unconsolidated material resulting in the formation of perceptible channels resulting in undulating terrain. The gullies are the first stage of excessive land dissection followed by their networking which leads to the development of ravinous land.

Ravinous Land : The word "Ravine" is usually associated not with an isolated gully but a network of gullies formed generally in deep alluvium and entering a nearby river, flowing much lower than the surrounding table-lands. The ravines, then, are extensive systems of gullies developed along river courses.

Undulating Upland with or without Scrub

This land is generally prone to degradation and may or may not have scrub cover. Such a land occupies topographically high locations in the respective systems. This excludes hilly and mountainous terrain.

Surface Waterlogged Land and Marsh

Surface waterlogged land is that land where the water is at/or near the surface and water stands for most of the year.

Marsh land is permanently or periodically inundated by water and is characterized by vegetation which includes grasses and reeds.

Salt-Affected Land

The salt-affected land is generally characterized as the land that has adverse effect on the growth of most of the plants due to the action or presence of excess soluble or excess exchangeable sodium.

Alkali land has an Exchangeable Sodium Percentage (ESP) of about 15 which is generally considered as the limit between normal and alkali soils. The predominant salts are carbonates and bicarbonates of sodium.

Shifting Cultivation Area

Such land is the result of cyclical land use consisting of felling of trees and burning of forest areas for growing crops. This results in extensive soil losses leading to land degradation.

Degraded Forest Land

Land as notified under the Forest Act and those lands with various types of forest cover, in which denudation of vegetative cover is less than 20 per cent of canopy cover, are classified as degraded land.

Degraded Pastures/Grazing Land

All those grazing lands in non-forest areas, whether or not they are permanent pastures or meadows, which have become degraded due to lack of proper soil conservation and drainage measures, fall under this category.

Degraded non-forest Plantation Land

Degraded lands containing plantation outside the notified forest area:

1. Barren rocky/stony wastes sheet rock area
2. Steep sloping areas:
   Land with very steep slopes (greater than 35 degrees) prone to erosion and mass wasting (landslides).
3. Snow covered and/or glacial area.

REFERENCES

Cochran, W.G. (1977). *Sampling Techniques* (3rd Edition), Wiley Eastern Ltd., (USA).

National Wasteland Development Board (1986). First meeting of the National Landuse and Wastelands Developmental Council (NLWC).

National Wasteland Development Board (1986). *Technical Task Group Report on Wastelands Definitions and Classification.*

National Remote Sensing Agency (1983). *Summary Report on Wasteland Mapping of India.*

National Remote Sensing Agency (1986). *Project Report on Mapping of Wastelands in India from Satellite Imagery 1980-82.*

11

Use of Remote Sensing Techniques for Mapping and Monitoring of Wastelands in India

N.C. Gautam, L.R.A. Narayan, G. Ch. Chennaiah, V. Raghav Swamy and *R. Nagaraja*

India is the seventh largest and the second most populous country in the world with unique physical and cultural diversity. The diversity in the physical landscape has resulted in different types of land which are subjected to different types of utilization to the maximum extent possible. Due to the increasing pressure of population, there is an excessive demand for more land, both for agricultural anal non-agricultural pursuits. This has resulted in the exploitation and reclamation of all such lands which are at present either lying unused or which have been turned unproductive due to various reasons. Hence, the immediate task is to have timely and accurate information, to the extent possible, on the location and extent of all such lands known as "wastelands".

The result of this study indicates that satellite data could be used for a broad wasteland development programme and bringing out the changes. The area under wastelands was calculated to form 5.33 lakhs sq.kms (16.21%) of the total geographical area of India. Eighty per cent to 90 per cent interpretation accuracy has been achieved when compared with ground surveys. The authors have also evolved a suitable wasteland classification system adaptable for using remote sensing techniques.

Study Area

India as a whole was taken for wasteland mapping – using remote sensing through visual interpretation techniques of Landsat MSS data.

Objectives

The main objective of this study was to utilize landsat satellites data for mapping and monitoring of wastelands and to evolve a standard classification system for mapping wastelands and also to quantify the area under each category of wasteland.

Limitations of Landsat Data and Mapping

1. Limited spatial (ground) resolution of the imagery, i.e. 1.1 acre or 0.45 ha.
2. The mappable unit on 1:1 million scale map being 1 sq.km (1 mm × 1 mm on the map) any isolated patch of less than 1 km of dimensions in any direction was not mapped.
3. Only broad wasteland categories were identified because of limited resolution of Landsat MSS data.
4. The spectral signature of two different objects may occasionally be similar and may pose problems for the interpreter to analyse the images properly. For example barren and fallow land have been kept together because of their bright tone and texture.
5. Keeping in mind some of the limitations of landsat data a more generalized system of classification was adopted based on the discussions between various people.

Nationwide Wasteland Classification System

Definitions

Wasteland is defined as that land which is degraded and is presently lying unutilized except as current fallows due to different constraints.

Classification System

Level I	*Level II*	
1. Culturable wasteland	1.1	Gullies or ravinous land
	1.2	Undulating upland with or without scrub
	1.3	Surface waterlogged land and marsh
	1.4	Salt-affected land
	1.5	Shifting cultivation area
	1.6	Degraded forest land
	1.7	Degraded pasture or grazing land
	1.8	Degraded non-forest plantation land
	1.9	Strip lands
	1.10	Sands
	1.11	Mining or industrial wasteland
2. Unculturable wasteland	2.1	Barren rocky/stony wastes/sheet rock areas
	2.2	Steep sloping area
	2.3	Snow covered or glacial area

Methodology for Wasteland Mapping

In the present study three types of data have been used, *viz.* Landsat data, secondary data and ground data.

Landsat Data

One hundred and ninety (190) landsat (MSS) false colour composites for the period of 1980-82 on 1:1 million scale covering the entire country were used for this study. Although, the scale of the images produced at NRSA varies from scene to scene slightly and the average scale is around 1:1.1 million, yet, the scale was corrected to 1:1 million while transferring the interpreted details on to the base maps using optical reflecting projector.

Secondary Data

Survey of India Topographical maps on 1:50,000 and 1:2,50,000 scale, Thematic maps produced by the National Atlas and Thematic Mapping Organization (NATMO) on 1:1 m, and maps of the Geological Survey of India (GSI) were used as main supporting data during both phases of interpretation and base map preparation.

In addition, existing information available on certain category of wasteland in the form of thematic maps of different scales along with other statistical data were also referred to and verified with the wasteland maps prepared from Landsat data.

Ground Data

Ground truth verification forms an important component in the methodology of interpretation of remote sense data. It could be in the form of sample verification on ground and collection of information about the different surface features on earth, for establishing correlations.

Analysis and Interpretation

Monoscopic visual interpretation technique based on the image characteristics such as colour, texture, pattern, shape, size, location and association enabled to identify and delineate both culturable and unculturable wastelands. Wherever, it was required Landsat images of Band 5 and 7 and False Colour Composite (FCC) of different seasons and years were also verified for better identification and delineation of wastelands. The analysis of work can be broadly grouped under three phases of activity, *viz.* preliminary analysis, ground truth and final interpretation of wastelands on Landsat imagery.

Preliminary Analysis

1. Selection of Landsat imagery of good quality and of appropriate season.
2. Understanding the advantages and limitations of Landsat data.
3. Establishing criteria and standardization of wasteland classification system.
4. Development of visual interpretation key based on image characteristics.

5. Interpretation of different categories of wasteland.
6. Identification of doubtful areas on the imagery.

Ground Truth

1. Physical verification of doubtful areas.
2. User interaction and collection of data.
3. Modification of wasteland boundaries.

Final Interpretation

Interpretation of different categories of wasteland has been carried out by visual interpretation key development on the basis of image characteristics (Table 11.1). Different wasteland classes (as mentioned in the classifications) were identified and delineated on the Landsat imagery. It may be mentioned here that this visual interpretation is subject to change depending upon the season, scale and resolution of the imagery. In the present study certain category of wastelands like salt-affected land, waterlogged marshy land, sandy area and snow covered or glacial area were easily delineated by virtue of their spectral separability, pattern and location, whereas gullied or ravinous land, shifting cultivation, etc., were delineated with moderate success. However, the category of undulating upland with or without scrubs, which is widely prevalent throughout the country, could not be easily delineated due to its merging with fallow land and similar spectral reflectance pattern. However, this was better identified and delineated using multidate imagery to the extent possible.

Table 11.1: Visual interpretation key

Sl.No.	*Wastelands*	*Image Characteristics*
1.	Salt-affected land	White to light blue tone of salt encrustations tonal variations subject to moisture content in the soil small size; irregular shape, fine to mottled texture; scattered and non-contiguous pattern; located near inland plains and coastal low lands; associated with irrigated land amidst

		crop areas and around tidal marshes. Season: January-March.
2.	Gullied or Ravinous Land	Light yellow to bluish green tone; tonal variation subject to moisture content and depth of the gully; varying in size and depth; irregular shape; coarse textures; linear and dendritic drainage pattern; located in plains along stream courses and on slopping grounds made of loose sediments; associated with severely eroded (water) areas, entrenched drainage. Season: January-March.
3.	Waterlogged or marshy land	Light to medium blue tone; tonal variation subject to water spread and organic content; varying in size; irregular in shape; fine to mottled texture; linear to scattered pattern; located in river plains and coastal low lands; associated with proximity to flood plains, coastal tidal flats and lagoons Season:December-April .
4.	Undulating upland with or without scrub	Light yellow to greenish blue tone; tonal variation subject to moisture content and vegetation cover; varying in size; irregular shape; fine to mottled texture; contiguous and scattered pattern; located on terrain with varying lithology; associated with gentle relief with moderate slopes; drainage & surrounding agricultural land. Season: August-October.
5.	Jhum or Forest Blank	Light yellow tone; tonal vegetation subject to vegetation cover; small in size; regular to irregular shape; fine to mottled texture; scattered pattern; located on hill slopes, associated with forest cover. Season: March-April.
6.	Sandy area (coastal or desert)	White to light tone; tonal variation subject to moisture content in the sand, varying in size; regular to irregular shape; fine texture; linear and continuous pattern; located in deserts; riverbed and coastal onshore; associated with shifting sand dunes; All seasons.
7.	Barren hill ridge or rock outcrop	Greenish blue tone; tonal variation subject to varying rock type; varying in size; irregular shape; fine to coarse texture; linear to contiguous and scattered pattern; located in plateau and hills, associated with barren and exposed rocky areas. All seasons
8.	Snow covered or glacial area	Bright white tone; tonal variation subject

to moisture content of dry snow or accumulated snow; large in size; irregular shape; fine texture; continuous pattern; located on mountain peaks and slopes; associated with high relief and glaciers. Season: December-March.

Transfer of Thematic Details

After interpreting wastelands on the landsat imagery, these were transferred on to a base map drawn on a transparent medium. In the base maps, drainage and other similar features provided the controls. Since the transferring of different wasteland categories onto the base maps was done manually, there may be at places mismatching of some details along the adjacent sheets. However, the mismatching of details along the adjacent sheets was rectified taking the drainage as control points.

Final Map Preparation

The thematic details were finally drawn on the base maps in different colours. The areas which are not included under wasteland were kept blank.

Results: Wasteland Pattern and Distribution

The distribution of wastelands suggests that it is more prevalent in the states of Andhra Pradesh, Bihar, Gujarat, Haryana, Karnataka, Madhya Pradesh, Maharashtra, Orissa, Punjab, Tamil Nadu, Uttar Pradesh and West Bengal. The aerial extent of wastelands is relatively less in other states of India. Several factors both natural and man-made can be attributed for the genesis and occurrence of wastelands in the country today. Besides, wind and water erosion, increasing salinity and alkalinity, increasing waterlogging conditions mainly due to canal irrigation and inadequate drainage, gully erosion, overgrazing, increasing deforestation and shifting cultivation, increasing process of desertification and other factors like non-availability water, presence of thick and extensive deposits of sand, etc., could be the contributing factors for the formation of wastelands.

The very high percentage of area under wastelands in States like Himachal Pradesh (36.67%), Jammu and Kashmir (60.14%), Sikkim (36.94%) and Arunachal Pradesh (14.48%) mainly due to a considerable area being under permanent snow and glacial cover whereas in the State of Rajasthan (37.81%) is because of desert sands and sand dunes.

Based on the present study the total area under wastelands in the country (with 1980-82 as base year) has been estimated around 53.3 million hec. or 16.21 per cent of the total geographical area (32.87 million hects.) of the country. Out of which 32.8 million hec. or 61.5 per cent is under culturable wasteland and 20.4 million hec. or 38.5 per cent under unculturable wasteland. This is a gross estimate based on a 1:1 m scale map.

Table 11.2 : Area under wasteland in India

		Category	Million hec.	
Culturable wasteland	1.	Salt-affected land	3.90	
	2.	Gullied or ravinous	4.32	
	3.	Waterlogged or marshy land	0.89	
	4.	Undulating upland with or without scrub	10.79	
	5.	Jhum or forest blank	2.40	
	6.	Sandy area	10.53	
		Total	32.83	
Unculturable wasteland	1.	Barren Hill ridge or Rock outcrop	2.75	Gross total 53.28 million hects.
	2.	Snow covered	17.70	
		Total	20.45	

Accuracy Estimation

In the present study the interpretation accuracy has been determined based on random sampling, wherein minimum number of points selected for each of the wasteland class was tested against known ground sample points and corroborated by previous knowledge of the area under study. Here, the interpretation accuracy has been calculated by applying the following formula:

$$P \pm 1.96 \quad \frac{(100 - P)}{n} + \frac{50}{n}$$

(After Snedecor and Cochran, 1967)

where P = desired proportion of estimated accuracy expressed in per cent

1.96 = is a constant (Z value assumed at 5% level of significance) applicable for accuracy estimated or remotely sensed data

For the present study 'P' is estimated at 85 per cent for all classes of wasteland. And the value of 'n' in minimum number of points (sample size) to be selected to estimate the interpretation accuracy has been derived by applying the following formula:

$$n + \frac{p \times q}{d^2}$$ (After Grejory, 1963)

where p = desired proportion or estimation accuracy

q = 100-p

d = standard error (5%) acceptable at 95% confidence level as determined from the formula

wherein p, q and d are expressed in per cent

For example:

p = 85%, q = 100-85 = 15% and d = 5%

$$n = \frac{85 \times 15}{5^2} = \frac{85 \times 15}{25} = 51$$

Hence n = 51

Therefore, interpretation accuracy at p = 85% and n = 51 will be

$$85\% \pm 1.96 \quad \frac{85 \times 15}{51} + \frac{50}{51}$$

85% ± 10.8%

85% ± 10.8% = 95.8% or

85% - 10.8% = 74.2%

Hence the interpretation accuracy ranges from 74.2% to 95.8%.

Conclusion

The present study demonstrates that satellite remote sensing is a valuable technique to map and monitor wastelands occurring over large areas, rapidly and economically. Thereby, providing a consistent and more accurate baseline information than many of the conventional surveys employed for such a task. The study has also brought out that the total area estimated under different types of wastelands occurring in the country is (53.3 million ha) higher than the total area reported (36.9 million ha) for the same year (1980-81) by the Directorate of Economics and Statistics, New Delhi. Uptodate information on location and spatial extent of wastelands is an essential prerequisite for the development of wastelands. This information when used with the information on other natural resources like soil, water, geology, etc., will help in the optimum utilization of wastelands either for agriculture, afforestation/plantation or postures, thus making it more productive.

12

Wasteland Development of Mewat Region of Haryana, its Identification and Planning

J. S. Yadav

The biggest challenge of the campaign for productivity lies in the agriculture sector, or, to be more precise, in respect of use of the land lying waste. It is a hidden resource which can be put to many productive uses. But it certainly requires appropriate identification of the area with the latest application of science and technology. For any developmental planning, updated resources data is very essential so that relevant strategy may be developed for its optimal use.

Mewat region is essentially dependent on agricultural production. To know the potentials available in the hidden wasteland, it was deemed desirable to know about the geology, geomorphology and groundwater conditions in the region in an updated manner, along with data on other aspects such as soil, land use and vegetation for integrated development. In this, remote sensing could play an effective role in collecting updated information. For this task, the Mewat Development Board approached the Indian Institute to Remote Sensing (IIRS), NSRA, Dehra Dun, through the Department of Industries, Government of Haryana, Chandigarh. The study was conducted during the year 1985. The present chapter is based upon the findings and recommendations of the study.

Mewat consists of five NES Blocks of Gurgaon district of Haryana, namely, Nuh, Ferozepur Jhirka, Tauru, Nagina and Punhana and one block Hathin of Faridabad district. It is bounded by Rajasthan from West and South, Delhi lies eastwards and the river Yamuna separates it from U.P. It is between longitude E78°35' and latitude N 27°31' to 28°21'. It has about 1874 sq. kms of area, it is known for its habitants,

Meos, a community name. The area is economically backward in comparison to the other areas of Haryana. Some sociologists call the Meos of Mewat "deemed Tribes".

Mewat has a varied topography. The valley has alluvial plains, several lakes and a small hill range, a part of the Aravalli runs on the one side in the entire area. Some parts are prone to floods. The soil of Mewat varies from sandy to loamy and certain low-lying parts having clayey soil have the problem of salinity. The climate of this region is more temperate than other parts of Haryana.

The land use statistics of Mewat area are given below—

(figs. in thousand hects)

1.	Total Area according to village paper	182
2.	Forest	01
3.	Land put to non-agri. uses	20
4.	Barren and uncultivable land	3
5.	Total of 3 and 4	23
6.	Permanent pastures and other grazing ground	1
7.	Land under misc. tree crops and groves not included in net area sown	nil
8.	Culturable waste	nil
9.	Total 6, 7 and 8	1
10.	Fallow lands other than current fallows	nil
11.	Current fallows	10
12.	Total 10 and 11	10
13.	Net are sown	147
14.	Culturable area (9 + 12 + 13)	158

Out of above classification of land, three main criterion can be taken up for identification of wasteland:

(1) Total Geographical Area (Net Area sown–Forest)

(2) Culturable Area–Net Area sown

(3) Barren and Uncultivable + Pastures and Grazing Grounds + Culturable Waste + Fallow Land.

Since no separate development plan, for wasteland has been made so far, no criteria have been put to practice for proper identification.

Stratification of Wasteland of Mewat

The total area has been classified under four major groups depending upon the processes and agencies responsible for their origins, namely, fluvial, aeolian, denudational and structural. They have been studied in detail and some of the main geomorphic features are given below:

Fluvial Land Forms

Fluvial processes play a very important role in the evaluation of landforms resulting from fluvial processes belong to constructional phases. In the Mewat, fluvial geomorphological units are as follows:

Alluvial Plain

Vast area of alluvial plain have been demarcated in the study area between the two major structural hills in the western part of the area, namely, Nuh Ferozepur Jhirka valley. The boundary and buried channels have been drawn from aerial photographs. The following wasteland forms have been identified in this valley:

(a) *Paleochannel*: When the photographs were analysed in the fields they were found to be narrowed linear depressions now almost entirely filled with sandy loams., At places water is present in the depression. But most of the area is under brackish water. To fight with this challenge, the Mewat Development Board (MDB) has taken up a scheme for the treatment of brackish water, on a pilot basis.

(b) *Waterlogged Area*: Old alluvial plain is almost flat and has many depressions which get filled with water during the rainy season and the dark tone of the picture shows the presence of water for long periods.

(c) *Old Natural Levels*: Old natural levels are seen in the vicinity of village Saundad. The profle configuration has almost entirely been obliterated by intensive agricultural land use. However, the remnant flood characteristics (i.e. slough near the natural levee. At present there is no water but it possesses dark tone due to high moisture content and is seen as depression) help in their recognition. MDB is taking up a scheme to meet this challenge.

Salt-Affected Area

The study area faces a severe problem of salinity. In total about 6,800 hectares are salt-affected in the Mewat region. It was very easy in field to identify this area with the aid of encrustation of salt on soil. The identification of sand sheet and the salt-affected area in the aerial photographs is slightly different because it also gives a dark tone.

Under DRDA and CADA programmes, gypsum is supplied to the farmers at 75 per cent subsidized rates.

Closed Basin

In the south-east of the area, near Biwan and Pahari, a close basin has been demarcated. This area is partially enclosed by isolated hills consisting of Ajabgarh series of rocks. One branch of Ajabgarh hills bifurcates and takes easterly turn and makes an accurate series of isolated hills, with an enclosed basin in between. Presently, there is no river to drain this area, the drainage seems to be choked in the area and hence more salt-affected patches are found. This vast depression encircled by irregular detached hill ranges is filled with salt and sand of alluvial origin. Lack of proper flushing, of surface water, introduction of surface canal irrigation system and excessive evapotranspiration due to increasing aridity might have contributed for the salt concentration.

Piedmont Zone

Piedmont Zone, as the name suggests, is a feature usually formed at the foot of a mountain. The term has also been used for the slopping land formed due to coalescence of alluvial fans, depositional as well as erosional in nature, locally covered with sand which together gives an undulating topography at the foot of the Alwar series. It consists of valley fills and fans. The materials which make the fan are coarse, transported material mixed with finer alluvium and at places sand deposits. The well sections at Sidhrawat show that these valley fills are of about 20 m. thickness. This zone starts from extreme south of Alwar hill ranges and extends up to Ferozepur Jhirka. From Ferozepur Jhirka, the sandy plain and active sand dunes have broken the continuity which again starts and extends up to the extreme north of the sandy area.

Aeolian Land Form

Sand Sheet

Some patches of sand have been noticed in the old flood plain. Since they are scattered throughout the old flood plain, without any particular fashion, it is difficult to state, about the origin of these sand sheets. These are thin at some places and thick at others.

Sand Dunes/Interdunal Depression Complex

The term interdunal depression complex is used here because it was not possible to demarcate the boundary of individual sand dunes either on photographs or in field owing to the small size of the dunes. West of the Nangli Mubarakpur village on the eastern side of the main range of Alwar, an area of about 2.5 km in length and 1 to 1.5 km in width has been demarcated as sand dunes/interdunal depression complex. During as storm, the sand spreads over nearby areas and sometimes forms sand dunes near the obstacles in cultivated areas.

Aeolian Sandy Plains

In Mewat a large area is covered by aeolian sandy plains in the western side of the region. The colour of the sand is yellowish brown. Here, sand is mixed with a proportion of silt and clay. The essential characteristic of this geomorphic unit is that it is located at the down slope of obstacle sand dune mounds. On the east of Alwar hills is the vicinity of village Nangli Mubarakpur, a narrow stretch of about 1 to 2 km in length has been demarcated. An extensive area is under a sand cover which is under seasonal cultivation at some places, otherwise the whole land is lying waste. This sandy plain with undulating topography suffers from sheet erosion. During the period of torrential rainfall running water from structural hills and from dissected sandy plains comes down and erodes in a sheet form covering a vast area which removes the human from the top soil. The problem of sheet erosion is related with the problem of gully erosion.

Since there was a great need to check these gullies, the Mewat Development Board has taken up a scheme for "Gully Plugging" being operationalized under the Department of Soil Conservation. It has given very good results and, at many places, a sizeable area has come under cultivation which has boosted the income of the farmers.

Denudational Forms

The Mewat region encompasses a variety of land forms produced by fluvial, aeolian and denudational processes. The geomorphic forms produced by the denudational processes have been discussed in this section. The following units have been delineated in the study area:

Pediment

This zone is one type of that group of erosional surfaces which is loping plain sculptured out or hard rock. In the Mewat area a few small pediments have been demarcated. These are shallow buried pediments with detrital cover and partially (shallow) weathered zone. In the southern Mewat area pediments are found between denuded hills and plains area, east of Ajabgarh rocks. It is found that pediment are of complex surfaces, comprising patches of partially weathered bedrock consisting of phyllite schists and slate.

The Government of Haryana should evolve a strategy for using the resources available properly. Some profitable and economically viable minerals can be exploited.

Buried Pediment

A pediment concealed under the thick and weathered mantle is found on the eastern side of the Ajabgarh rock. The eastern boundary of buried pediments is conjecturally drawn with the help of delineation of denudational hill and pediments. The western side of the buried pediment zone is flanked by sandy plain and sand dunes.

Inselberg

Isolated denudational hill rocks as residual out crops have been found in this area and termed as Inselberg. An inselberg is in the literal sense as upland hill rock and its essential feature is its abrupt rise from the surrounding plains, where the latter may be of the erosional or depositional type. The term inselberg encompasses a considerable range of morphological type.

Structural Hills

Two main types of structural hills have been demarcated on the basis of the lithology, *viz.* Alwar hills and Ajabgarh hills. The general trend, of these hills is NE to SW. A few isolated hills erosionally detached from the main range but prominently exhibiting the structural control over the topography have been included in the structural hill unit.

The present unit forms the highest topography in the area reaching to a maximum height 750 metres above MSL. Structural hills are characterized by hogback to cuesta type of land forms showing folding, faulting, dragging, jointing, and overturning of beds in them, subsequently modified by erosional agencies. Debris and rockfall have been noticed in adjoining scarpedges. Due to quarrying a good deal of material is deposited in the form of scree. Some landslides are also found in the vicinity of quarries and along steep slopes. It is difficult to separate the scree deposits formed by human activities and those formed by normal gravity forge. Some structural hills show flat tops which are possibly remnants of the earlier development plantation surfaces.

Mewat Development Board and Wasteland Development Department

Mewat Development Board (MDB) realized that the absence of vegetation means that above the ground level there is nothing to reduce the wind velocity and below the ground there is no root system to blind the soil particles together, as in addition there is no soil moisture to give coherence to the particles. As a result, wind is able to play a more significant role in damaging the environment. So, it was felt necessary that large-scale plantation be raised on wastelands owned by the Panchayats and the Village communities of this area. Additional funds are being released from out of the MDB funds to the Forest Department to undertake intensively the various schemes like afforestation on panchayat/wastelands, farm forestry, raising of shelter belts, afforestation on saline and alkaline soils, etc. So far a sum of Rs. 78.89 lakhs has been spent and forests have been developed on 1,336 hects. and 1,719 RKMs. up to the year 1985-86.

Soil Conservation and Agriculture

Keeping in view the low agricultural production owing to various factors like undulating topography, lack of adequate irrigation facilities, etc., the Department of Agriculture, Haryana, has been involved in a big way in the proper use of wasteland in Mewat. In addition to the normal schemes being implemented by the Agriculture Department, the Mewat Development Board has also approved the following schemes:

Watershed Management

Due to presence of the Aravali range, the area has a very undulating topography as discussed above, thereby creating problems of soil erosion during the rainy season. Ninety villages located on the sides of the Aravali ranges are adversely affected due to soil erosion and the affected area is about 3,639 hects., which goes waste every year. The area has been split into 90 mini-watersheds. The treatment of each mini-watershed is being taken in a phased manner in a period of five years. An integrated project of watershed management with an outlay of Rs. 126.66 lakhs has been prepared which includes schemes like soil conservation works, afforestation, fisheries, horticulture, etc. The scheme was taken up for implementation by the Soil Conservation Department for construction of embankments/drought ponds for arresting run-off water, reclamation of gullies and prevention of gully formation, land treatment to check erosion, etc. So far, a sum of Rs. 88.03 lakhs has been spent and an area of 1,368 hects. has been plugged and 81 percolation embankments have been made.

Conclusion

On the basis of the above analysis, we may conclude that for an identification of wasteland, we must adopt the latest scientific techniques developed, like remote sensing besides Revenue Record exercise. Application of science and technology will give us more information regarding putting the wastelands to manifold uses along with their identification. The results obtained through remote sensing would be more reliable from policy and planning point of view. So, the criteria for identification of wastelands should be such as offers a way to successfully solve the problem. Until and unless identification of wasteland is associated with the identification of its remedies for various uses through the new resources available here itself, the exercise will be futile.

In the case of Mewat wasteland development following policy implications are put forward:

Policy Implications

(1) For structural hill development the Government should lay more emphasis on afforestation to maintain ecological balance. The hills should be put to micro-testing, blasting, shooting experiments, etc., and more cantonmental concentration be given. More stone quarries should be established. Another important mineral available in the area of the Nuh and Hathin Block is "Salt Peter" which is commonly known as potassium nitrate. The Government should take steps for surveying the area where Salt Peter is available to find out the exact quantity available, its quality and feasibility for its commercial exploitation for similar industrial use and setting up fireworks manufacturing and other Salt Peter based new industries.

(2) In the south-east of Mewat near Biwan, drainage seems to be choked in the area and hence more salt-affected patches are found. This should be controlled immediately otherwise it may endanger more soil.

(3) More emphasis is required on soil conservation works like construction of percolation embankments, water harvesting tanks, gully plugging, conservation of moisture, etc., for the improvement of piedmont zone.

13

Wasteland Reclamation and Development in Uttar Pradesh

P. D. Bajpai

For the ever-increasing population in the State of U.P., demand for land is also simultaneously increasing. Therefore, any effort made for reclaiming unutilised areas for useful purposes for growing crops and trees is worthy of money to be spent on this account. Unused and unproductive land occupies a large area in the State, i.e. lands infested with shrubs and bushes, ravine land, waterlogged lands, saline and alkaline lands, riverine lands, stony and gravelly lands and thin cover lands of Bundelkhand region and hilly tracts.

Causes and Extent of Land Deterioration

There are several reasons due to which once good and fertile and cultivated lands have deteriorated to such an extent that they are presently not being utilized. Most important amongst the reasons for the deterioration is the misuse and undue interference by men and animals resulting in a disturbance of the ecological balance beyond critical limits.

The method and treatment vary according to the category of land and the extent of deterioration, cost of development, the nature of deterioration and types of lands. The benefit-cost ratio determines the economic viability or otherwise of reclamation purposes. These deteriorated lands by and large belong to the poor farmers who are incapable of investing the money required for the improvement of these lands. Because of this reason, the scant attention has been paid to this problem. Therefore, the deterioration which started at a slow rate in the start has assumed an enormous magnitude. In the long process of neglect, some of these affected areas have become convenient hide-outs for anti-social elements (dacoits) leading to a menace and great nuisance to society.

The extent of area under different categories of wasteland is not precisely available. However, on the basis of sketchy information gathered from revenue records and other relevant sources, estimated area under different categories and deteriorated land is given in Table 13.1.

Table 13.1 : Area under different categories of wasteland in U.P.

Sl.No.	*Category of wastelands iri U.P.*	*Area in lakh hectares*
1.	Land infested with shrubs and bushes	N.A.
2.	Ravine area	12.30
3.	Waterlogged lands	18.70
4.	Saline and alkaline lands	11.50
5.	Riverine lands	15.00
6.	Stony and gravelly lands	N.A.
7.	Thin cover lands	N.A.

It is thus evident that out of the total reporting area of the State, a high percentage is under wastelands.

Magnitude and Causes of Deterioration and Methods of Reclamation

Land Infested with Shrubs and Bushes

Since land infested with shrubs and bushes has not been differentiated in land utilization statistics, reliable data with respect to this category are not available. These areas are a part of wastelands and are large in U.P.

There are several reasons for areas not being utilized, mainly the presence of deep-root grasses and shrubs, shallow soils with low fertility, severe erosion, etc. In the past, some of these lands have beer. reclaimed, but most of the areas still need improvement. Tractorization on these difficult lands is a costly venture leading to unfavourable benefit-cost ratio at the beginning but in the long run, these lands can be economically viable.

It is important in the first phase to survey these soils in order to delineate different land capability classes indicating suitability of such lands for crops, grasses and/or forests. Most of the area, however, may be put under fodder and fuelwood development. The National Commission on Agriculture (1976) recommended that after suitably developing these lands, they may be reserved for fodder and linked up with an intensive programme of cattle and sheep development.

Work of the development of these lands may be assigned to the Land Development Corporation with adequate financial assistance.

The Ravines

Some lands along the banks of rivers like the Yamuna, Chambal and others with their tributaries have been extensively eroded, ultimately resulting in shallow and deep gullies, commonly known as ravines. The rivers in high floods back up to ravines and hasten soil erosion resulting in removal of fertile soils from the upper tracts (table-lands). The process of gully formation once started continues with unceasing speed leading to higher coverage of area under ravines.

Reclamation of ravine lands is urgent because of its fast spread into cultivated lands and also because of the presence of dacoits in the ravines which provide escape routes and hide-outs for such elements. Besides, there is a huge economic loss of production to the extent of about a few million tonnes of food grains annually.

Ravines of U.P. have been classified into very small and small gullies, medium gullies and deep and narrow gullies as given in Table 13.2.

Table 13.2: Classification of ravines of U.P.

Sl. No.	*Types of gullies*	*Specification*		
		Depth (metres)	*Bed width (metres)*	*Side slope (%)*
1.	Very small and small gullies	–	18	Varies
2.	Medium gullies	< 3	18	8-15
3.	Deep and narrow gullies	(a) 3-9 (b) 7-9	18	Varies, mostly steep

Reclamation work may be economical and suitable in areas with small and medium gullies. In the case of deep gullies cost of reclamation may be prohibitory because of very large earth work. Table-lands of deep ravines are reclaimed for crops. Deeper areas and sloppy ravines are used for forest and fodder development. Suitable outlets are provided to drain out excess run-off; check dams and gully plugs are provided, wherever necessary. Sometimes, horticultural crops are produced with suitable provision of contour bunds of necessary grades.

According to rough estimates, about 1.23 million hectares is ravinous land in U.P. Out of this area, only 41,881 hectares have so far been reclaimed and this reclaimed area has been developed in about 17 years.

Presently, a scheme is in operation in three districts, namely, Kanpur, Etawah and Agra. In each year about 3,000 hectares are being treated at present. It is evident that the rate of reclamation is slow to have any impact, and, at the present rate, it will take a very very long time to reclaim this large area.

The cost of reclamation of ravinous lands varies according to the item of work to be undertaken depending upon whether it is done mechanically or manually. In most of the cases reclamation of such lands is economical. However, economic benefits should not be the criteria in reclamation of ravines. The reclamation is important for social upliftment in way of removing the poverty in the area and to plug the hide-outs of anti-social elements. It is also of immediate necessity to stop the spread of ravines.

With the present rate of reclamation of ravinous lands, it may take a long period before all such lands are taken under cultivation. It is of paramount importance, therefore, that a full-fledged agency should be entrusted the job with necessary financial resources in the form of centrally as well as State-sponsored schemes.

It is also necessary to properly survey the whole ravinous area in the State, categorize and classify these lands according to land capability classes so that its reclamation may be planned in a scientific manner.

Waterlogged lands

Reliable data on the extent of waterlogged area are not available and the area indicated at waterlogged is approximate. According to Draft Annual Plan (1986-87) of the Agriculture Department (Soil Conservation Wing) about 1.87 million hectare area is waterlogged and flooded. By constructing roads, railways, canals and also building under new townships, the natural drainage system has been dislocated and run-off of rains is obstructed. Seepage from canals, irrigated cultivation without drains and excess use of irrigated water contribute substantially to waterlogging.

In waterlogged areas, the water-table rises to an extent that soil pores in a root zone of a crop become saturated restricting air supply. The actual depth of the water-table, when it starts affecting crop growth adversely, may vary widely from zero for crops like rice to about 1.5 metres for other crops. Wheat and sugarcane are affected when the water-table is within 0.6 metre. Maize, bajra and cotton are sensitive within 1.2 metres and gram and barley to within 0.9 metre. It is desirable to maintain the water-table below the capillary range, say, minimum around five metres. Waterlogged conditions not only retard crop growth but also increase salt concentration at the surface leading ultimately to salinity and alkalinity problems, besides deteriorating physical conditions of soils.

Reclamation of waterlogged areas involves development of a proper drainage system surface or sub-surface or both, lining of canals to avoid excessive seepage, sinking augmenting tubewells and execution of proper water management at farmer's level. Proper crop sequence and suitable agronomic practices may also minimize adverse affects of waterlogging. Afforestation of these areas may be of considerable economic importance. Planting trees on ridges or mounts may be taken on a large scale. These lands can also be utilized for growing grasses like Para, Dallis, Napier and Guinea.

Proper survey, planning and execution of the programme of reclamation should be done and the efforts of the departments of irrigation, forest and agriculture should be coordinated. In fact, a coordinating agency should be created so that these unused areas rapidly developed and deterioration of soils avoided in future.

Saline and Alkali Soils

Saline and alkali soils are characterized by the presence of excessive salt concentration which may occur due to various reasons like (a) Capillary rise from subsoil, (b) Excessive and indiscriminate use of canal water, (c) Accumulation of salts brought down by river water plains and subsequent deposition with alluvial materials, (d) Seepage from canal, (e) Mismanagement of lands and (f) Faulty drainage system. Saline soils are characterized by the presence of excess salts like sodium chloride and sulphate, but very little of sodium corbonate and sodium bicarbonate while in the case of alkali soils sodium carbonates and sodium bicarbonates are in excess with very little of sodium chloride and sulphate.

In U.P., about 1.5 million hectares of land is affected by salinity and alkalinity problems. Thirty-six districts of the plains of U.P. are affected with excessive salt concentration. Most of the affected soils are alkali in nature while some areas in the districts of Agra, Mathura and Aligarh are affected by salinity. A list showing the districts affected with salinity and alkalinity problems along with the area affected is given at Annexure : 13.1.

For classes of saline-alkali are recognized and can be differentiated by the magnitude of electrical conductivity measure of salinity, Exchangeable Sodium Percentage (ESP)—a measure alkalinity and pH measure of alkalinity (also acidity). Their distinguishing characteristics are given in Annexure : 13.2.

The problems of reclamation are much more complicated under field conditions mainly because of the high cost involved in reclamation and poverty of cultivators of these lands. Besides, an area approach for creating facilities for irrigation and drains is required for which cooperation of other farmers and coordination of the revenue staff is a prerequisite but mostly it is lacking.

Distribution and application of soil amendments like pyrite and gypsum alone do not reclaim the really Usar lands for good. They have to be tackled in an integrated manner after provision of assured irrigation and drainage and has to be supplemented with proper crop sequence, past reclamation agronomy and constant watch and care to prevent the soils from reverting back to their original condition.

Because of these shortcomings, no large scale reclamation projects have been initiated so far, even though reclamation techniques are known.

In the State, there are about 1.15 million hectares of salt-affected lands spread over 36 plains districts. Besides, there are still more large areas which are marginally alkali soils and are mild or moderate in nature. However, these lands have not been surveyed and categorized though under the Department of Agriculture two schemes are, at present, in operation:

(1) Reclamation of Alkali Lands in U.P.

This scheme is running since 1977-78 and so far 3,51,366 MT of soil amendments have been distributed and it is claimed that 1,04,737 hectare area has been reclaimed. But only mild alkali lands are being reclaimed under the scheme. However, no part of the 1.15 million ha. affected area has been reclaimed, leaving whole of this area almost unreclaimed.

(2) Reclamation of Usar Lands of Allottees

This scheme was sanctioned for districts of Lucknow, Moradabad, Etah, Fatehpur and Unnao. An integrated programme of usar reclamation including on-farm development, provision of irrigational water and drainage besides use of soil amendments is being implemented for the benefit of new allottees of lands who are mostly Scheduled Castes. So far about 1,766 ha land have been reclaimed.

Besides the above two schemes, the U.P. Bhumi Sudhar Nigam has reclaimed about 152 ha Usar lands in the district of Lucknow. In addition in Usar reclamation programme in 10 factory areas in nine districts is also in operation. So far about 7,000 ha land have been treated with application of about 42,000 MT of soil amendments.

Efforts so far made in reclamation of these soils are meagre in view of the extensive area of about 1.2 million ha affected by excess salt accumulation. Reclamation work has to be taken up on a war footing in an integrated manner which includes not only application of soil amendments, but also on-farm development, provision of irrigation and drainage and constant watch on post-reclamation agronomy. It is also desirable that salt-affected soils should be surveyed, classified and mapped for proper planning and execution of reclamation programmes. The U.P. Bhumi Sudhar Nigam should be strengthened with financial assistance. Central and State projects on Usar reclamation should be allowed to run by the Nigam.

Riverine Land

The heavy sediment load carried by rivers in India and steep slopes they have to negotiate cause meandering action and serious bank erosion. This is specially so with the rivers originating in Himalayas. Fertile areas with flourishing crops, orchards, towns and cities on the banks of the rivers are often eroded by meandering streams. Depending on the course of the rivers and the speed of flow, they deposit large quantities of sediments on their way to sea. Gradually, landmasses are formed which remain for some part of the season, at least after the rainy season, above the flood level. Such riverine lands are called "Khader" or "Diara". It has been estimated that about 1.5 million ha of such lands occur in U.P. These lands run along the rive course but their width varies with the flow characteristics of the rivers. The areas under riverine lands can be reclaimed by canalizing the river flow. The Sutlej, for instance, has a wide stream bed of 8 to 10 km in width. During 1962-65, embankments were constructed and the Sutlej, in this vulnerable stretch, has been canalized to confine its flow to a narrow channel, reclaiming thereby 1,00,000 ha for cultivation. Another example is the Beas riverine land. Here the flood embankment has been constructed, reclaiming thereby about 25,000 ha of riverine wasteland for cultivation. Constructing similar embankments on vulnerable stretches of the rivers of Ravi and Jamuna, 25,000 to 30,000 ha of riverine wasteland can be brought back to cultivation. Other examples are Patiala Ki 'Rao' and Janta Devi Ki "Rao". They used to flow through the districts of Ambala, Patiala and Sangrur and then join the Ghaggar. These "Raos" have since been diverted into the Sultej via Siswan. In consequence, large areas of land formerly lying under water on both sides of these "Raos" and also the stream bed are gradually being brought under cultivation.

In view of the above facts, it is necessary that in the multiple channel rivers occupying vast areas under their beds, the subsidiary channels should be diverted, wherever feasible, into the main river and the land under the bed reclaimed for profitable land use.

It is further suggested that rivers on which dams have been constructed to effect flood moderation, the canalization of such rivers may be effected downstream of the dam, wherever feasible, in bringing back large areas of riverine wastelands for profitable land use.

It is also possible to utilize these "Diara" (riverine) lands by adopting suitable crops and agronomic practices. Some of the improved techniques have been developed, the cultivators should be motivated to adopt them.

A huge sediment load is carried by hill torrents called "Chos" emanating from the Shiwaliks. These sediments get deposited when a "Cho" emerges in the plains where it divides and subdivides and ultimately disappears after about 20-25 km from the foothills. In the course of such action "Chos" deposit thick blankets of coarse debris, sand, etc., over the green fields. It is suggested that, wherever feasible, techniques and experiences already available for grouping "Chos" (hill torrents) for their canalization and reclaiming the "Cho" devastated land for better land use should be adopted extensively.

Stony and Gravelly Lands

No data on the area under stony and gravelly land are available. However, a large area of such land exists in all eight hill districts as well as in the Bundelkhand region. The lands are mostly found on rolling topography which have changed them to wasteland, also because of heavy grazing and indiscriminate cutting of trees, and bushes, because of inadequate vegetation, they are subjected to soil erosion. In many cases because of salt removal to a high degree, such lands become acidic in nature. These soils can be reclaimed by liming, tree plantation and restricted grazing.

High Altitude, Steep Slopes and Meadows

No data on the extent of high altitude, steep slopes and meadows are available, but it is an extensive area in the Himalayan region. In the case of injudicious land use, soil erosion takes place to a very great extent. Heavy silt removal not only leaves the spot naked devoid of herbs and plants, but also endangers the life of multipurpose like growing of suitable trees and grasses. So far no attempt has been made towards this end mainly because of inaccessibility to this difficult terrain. Valleys in high altitudes with a peaty soil have not at all been exploited; urgent attention is also required for increasing production from pastures in these areas.

Besides above wastelands falling in categories of waterlogged lands, ravine lands, saline and alkali lands, lands infested with shrubs and bushes, riverine lands, stony and gravelly lands and high altitudes, slopes and meadows, lands lying waste along railway lines in the State for which no data are available because of lack of suitable records. Similarly, extensive areas lying waste or unutilized or underutilized along National and State highways, rivers, canals, etc., need be properly surveyed and after reclamation, wherever necessary, utilized according to Land Capability classes. It requires some organized action with the coordination of concerned departments.

ANNEXURE : 13.1

District-wise Area of Usar Land in U.P.

Sl.No.	*Name of district*	*Area in ha*
1.	Azamgarh	32045
	Total: Gorakhpur Division:	32045
2.	Lucknow	25515
3.	Unnao	48638
4.	Rae Barelly	59800
5.	Sitapur	19932
6.	Hardoi	35535
7.	Lakhimpur-kheri	24083
	Total : Lucknow Division:	213503
8.	Faizabad	26599
9.	Sultanpur	42577
10.	Pratapgarh	31685
11.	Barabanki	23966
	Total : Faizabad Division:	124827

Sl.No.	Name of district	Area in ha
12.	Saharanpur	12022
13.	Muzaffarnagar	19763
14.	Meerut	15748
15.	Ghaziabad	20439
16.	Bulandshahar	32345
	Total: Meerut Division:	100317
17.	Aligarh	45057
18.	Mathura	12563
19.	Agra	25049
20.	Mainpuri	66752
21.	Etah	56027
	Total: Agra Division:	205440
22.	Farrukhabad	49499
23.	Etawah	42328
24.	Kanpur	78123
25.	Fatehpur	41937
26.	Allahabad	68232
	Total: Allahabad Division:	280124
27.	Bareilly	18588
28.	Badaun	25283
29.	Moradabad	28456
30.	Shahjahanpur	21725
	Total: Bareilly Division:	94052
31.	Varanasi	23328
32.	Jaunpur	28456
33.	Ghazipur	14796
34.	Ballia	21218
	Total: Varanasi Division:	88227
	GRAND TOTAL:	1138543 or 1139000

ANNEXURE : 13.2

Distinguishing Characteristics of Saline Alkali Lands

Class Name	*Electrical conductivity (Ecx10³-millionhs/cm.)*	*Exchangeable sodium Percentage (ESP)*	*pH*	*Other features*	*Reclamation requirements*
1. Saline	> 4.0	< 15	< 8.5	Predominant anions–chlorides sulphates, a little of bicarbonates and nitrates; good aggregation water permeability	Leaching of salts and improvement of drainage.
2. Saline-alkali	> 4.0	>15	> 8.5	Predominant anions–chlorides, sulphates, bicarbonates and carbonates	Pyrite and gypsum are most suitable amendments.
3. Non-saline-alkali	< 4.0	> 15	> 8.5	Predominant anions–carbonates bicarbonates associated with sodium–dispersed clay, very low water permeability.	Same as above in case of saline-alkali.
4. Degraded alkali	< 4.2	> 15	< 7.0	Exchange complex is dominated by hydrogen ion and nitrogen ion; salts almost completely removed–physical condition as of non-sallne alkali.	Very difficult to reclaim.

14

New Approach for Growing Eucalyptus on Alkali Soils

K.K. Mehta

In our country we have 11 per cent area under forests against the minimum 22 per cent. So, it is very essential to identify lands where more trees can be planted without adversely affecting our crop production. One of the prospective areas is alkali-affected areas locally known as "Usar" or "Kallar" lands. These lands are lying waste and cover about 3 million hectares mainly in the States of Haryana, Punjab and Uttar Pradesh. Many efforts have been made from time to time to grow trees on these lands, but not with much success except for Valaiti Babool (Prosopis) and Kikar (Acacia). In this chapter, the performance and approach for planting eucalyptus trees on alkali soils has been presented. Although other tree species can also be planted by adopting the following approach, this chapter's emphasis is eucalyptus because it is more suited to the farmers as its wood can be used for many purposes, finds ready market, leads to more income in comparatively less time and it does not create any bird problem for the farmers as the birds cannot make nests in these trees.

Existing Methods of Planting

Two methods of planting trees on alkali soils are generally recommended which aim at reclaiming only a little portion of the soil.

These are: (i) Pit method, and (ii) Augerhole method.

In the pit method pits of 1 m × 1 m size are dug. The pits are filled back with soil brought from same good field or with the same alkali soil plus gypsum and farmyard manure. This method cannot be adopted on a large-scale because it requires a lot of labour and money at the initial stages.

The augerhole method involves making holes of 15 cms diameter up to a depth of about 150 cms so as to break the hard "Kankar" layer which generally exists at a depth of about one metre. The holes are filled with alkali soil and gypsum plus FYM. This method could not become popular with the farmers because of the following difficulties:

(i) It is very difficult to make holes using augers manually especially during the months of April to June when the soil is very hard. It is almost impossible to break the "Kankar" layer by a labourer of medium health.

(ii) Even if the hole is made mechanically the growth rate of eucalyptus trees is reduced drastically after 2-3 years. The fact is that the roots grow only in the hole and do not grow side-ways as the soil outside the hole remains highly alkaline which the eucalyptus tree cannot tolerate. After 4-5 years the growth rate becomes almost zero.

New Approach

In this new approach, the results of which have also been presented in this chapter, involves reclamation of the whole alkali field, both at the surface and in the deeper layers to such an extent that the particular tree species is able to tolerate. The alkali field is bunded, levelled and gypsum at 10 to 15 t/ha is applied during the months of June-July. Rice/Wheat/Berseem crop rotation is followed for the first 4-5 years. In this way the soil is improved up to one metre depth to such a level that eucalyptus roots can grow in all the directions. The surface soil layer is enriched with organic matter and other plant nutrients. In addition to all these benefits, the farmer gets income by growing rice and wheat which not only covers the cost of reclamation but gets additional income. The experience on more than 400 farmers' fields during 1975-85 has shown that the farmers could get 4.5 t/ha of rice and 1.8 t/ha of wheat during the first year of reclamation. These yields increased to about 55 qtl/ha for rice and 25 qtl/ha for wheat within four years without further application of gypsum. At this stage, the farmer is advised to plant eucalyptus saplings in his semi-reclaimed alkali field without application of gypsum and without digging pits or making augerholes.

It is advised to plant 6-month-old saplings having about 89-100 cm height raised in polythene bags. If the full field is not to be planted under trees then trees can be planted only on the boundary bunds of the fields and near the tubewells. Older saplings are recommended as they are more tolerant to adverse conditions. These saplings should be planted during the end of June to August or during February-March. Before planting, upper soil up to 30 cm depth should be spade-loosened. At the time of planting, the polythene bags around the roots of the saplings should either be removed or the bottom of the bags should be scratched so that roots can grow freely. Aldrin should be used to protect the plants from white ants.

If planted in a block of field, the distance from row to row and plant to plant should be kept three metres. This will allow 1500 to 2000 plants per hectare. Denser plantation will lead to thin and long plants.

Care should be taken that only smaller doses of urea at 25 gm/plant should be applied at a time with an interval of one month. Higher doses of urea per plant can lead to mortality of the plants, especially in the initial stages.

In the winter season, Berseem should be sown in between eucalyptus plantings during the initial three years which will fetch an additional income and also improve the soil.

Experiences on Farmer's field

A 0.4 hectare field of S. Darshan Singh of village Kachhwa was lying waste as it was highly alkaline having pH_2 values of 10.4 and electrical conductivity values of 2.6 mm hos/cm in 1:2 soil : water suspension at the surface. In fact, the whole soil profile was highly alkaline and had excessive soluble salts (Table 14.1).

This land was levelled and treated with gypsum at 12.5 t/ha. During 1975-81, rice-wheat crop rotation was followed. The grain yields of rice and wheat were more than 50 and 20 qtl/ha respectively (Table 14.2).

Table 14.1: Improvement in soil properties on the farmer's alkali field during cultivation of crops and growth of eucalyptus

Depth (cms)	*pH_2*			*EC_2 mmhos/crn*		
	Original 1975	*Before eucalyptus plantation 1981*	*After 5 years of tree plant-ation 1986*	*Original 1975*	*Before Euc. plant. 1981*	*After 5 years of Euc. plant. 1986*
0-15	10.4	8.6	8.7	2.6	0.24	0.18
15-30	10.4	9.0	9.0	2.0	0.46	0.39
30-45	10.3	9.2	9.0	1.6	0.60	0.34
45-60	10.0	9.2	9.0	1.3	0.56	0.44
60-75	10.0	9.3	9.2	1.2	0.52	1.00
75-90	10.0	9.3	9.2	1.2	0.52	0.44
90-105	–	9.5	9.1	–	0.50	0.40
105-120	–	9.6	9.1	–	0.50	0.33

Table 14.2 : Grain yields of rice and wheat on farmers' alkali field before planting eucalyptus trees

Crop	*Year*					
	1975-76	*1976-77*	*1977-78*	*1978-79*	*1979-80*	*1980-81*
			qtl./ha			
Rice	42.0	50.0	55.0	59.5	60.0	59.0
Wheat	16.0	18.0	21.0	25.0	24.0	25.0

The soil also improved not only in the surface layer but also in the deeper layers. The pH was reduced from 10.4 to 8.6 and EC_2 values from 2.6 to 0.24 dsm^{-1} (Table 14.1). Water absorption capacity of the soil also increased. At this stage 600 eucalyptus saplings were planted on this field by following the above-mentioned approach. Growth of these trees was monitored by taking height and girth measurements every year during September-October. It has been observed that after four years of growth the average height of the plants was 13.0 metres and girth 47.0 cms. during October, 1985 (Fig. 14.1). The average rate of increase in height and girth was 3.0 metres and 9.5 cms per year. The growth rate is very good in comparison to the growth on normal soils. In fact, there are many side branches on every tree. The green colour of the leaves and a large number of side branches give an unchanting look to the whole field.

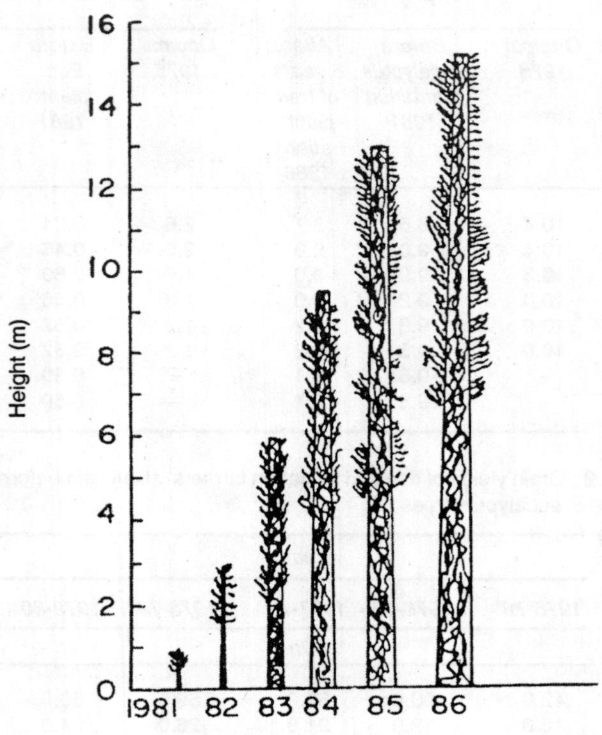

Fig. 14.1 : Girth in different years

Berseem in Eucalyptus Plantings

Berseem was also grown as intercrop (fodder) during winter seasons in the initial three years (1981-84) of eucalyptus growth on the above-mentioned field. The crop was so good that the farmer sold this berseem fodder for Rs. 2,000, Rs. 3,000 and Rs. 2,500 per ha during the first, second and third years respectively. Thus, on the one hand, the trees grew well and on the other, the farmer was able to get an additional income during the initial three years by growing Berseem in between eucalyptus plantings. This approach also helped in making the optimum use of the irrigation water. The applied water was not only used by Berseem but also by trees thus ensuring satisfactory growth of both trees and Berseem. The additional benefit of growing Berseem is that it is a leguminous crop which fixes atmospheric nitrogen in the soil and also adds organic matter due to falling of leaves and root stubbles. Berseem was not grown in the fourth year because the eucalyptus trees grew in height and the whole land was fully covered with vegetation. Moreover, the trees also required more support to save them from falling in strong winds. So, soil was put around the trunk of the trees.

Improvement in Soil Properties

During the growth of eucalyptus and Berseem soil has further improved. The pH levels have decreased further especially in the deeper layers. Similarly, the salt content has also decreased during 1981-85 which is evident from the decrease in the electrical conductivity of the soil (Table 14.1).

Impact on Farmers in the Surrounding Areas

Seeing the good growth of eucalyptus this farmer has planted eucalyptus on an additional 0.8 ha semi-reclaimed alkali soil. The farmers in the surrounding areas are also much impressed with the excellent growth of eucalyptus on this semi-reclaimed alkali field. Earlier they were thinking that rice, wheat or Berseem could be grown only on alkali soils. But now seeing the success of these trees more than one lakh eucalyptus saplings have been planted in village Kachhwa and in three other nearby villages, i.e. Sugga, Sambhli and Birnaraina where on the Operational Research Project for reclamation of alkali soils is being carried our.

Future Prospects

If this approach of planting trees in the semi-reclaimed alkali soils is adopted then a fraction of about 3.0 million hectares of alkali land which are being reclaimed will not only produce rice, wheat and Berseem but will also be put under trees to the benefit of the farmers. Planting of trees will also help the farmers to make the best use of the limited irrigation water resources. After establishment, the plants require very little less irrigation. It will also help the farmer to make the best use of his leisure time to look after the trees. Other tree species such as *acacia*, mulberry, *shisham*, *neem* (Azadiracta indica) have also been grown successfully by this approach in the semi-reclaimed soils but their number is still less in comparison to eucalyptus. It is expected that if this approach is adopted properly, more trees will be planted to the benefit of the farmers.

15

Levelling and Afforestation of one Million Hectare Dotty Land situated Along the Banks of Chambal River

S.S.S. Shastri

India, a sub-continent, with an area of 32,80,483 sq. km and a staggering population of one hundred crores by the end of this century, requires foodgrain to feed, clothing to cover, dwellings to shelter and fuel to keep it maintained, protected and mobiles and in order to secure a well synchronized growth of the country.

India is undergoing a Green Revolution under the 20-Point National Programme launched by our late Prime Minister, Mrs. Indira Gandhi. However, only 45.5 per cent area of the country out of total 32,80,483 sq. km is under cultivation and part of the remaining 55 per cent, i.e. 6.4 per cent is lying unused and uncultivated. Apart from this, the major portion of the land presently under cultivation, i.e. 45.5 per cent is in the extensive agriculture system whereas only a little fraction of the total cultivated land (45.5%) is in the intensive agriculture process.

The state of Indian agriculture is, however, predominantly of the subsistence type. Nearly three quarters of the total area under cultivation is devoted to production of food crops and yet the country is not self-sufficient even in its minimum food requirements. The 70 per cent of the Indian population engaged in agriculture is still unable to produce enough food to feed the remaining 30 per cent. The fact is that even with such a large proportion of people working in the fields and farms, the production is low as their efficiency is very low, i.e., only 8 per cent compared to a farm worker of the U.S.A. No doubt, our farm workers labour much harder than an average American farm labourer but without the aid of modern farming equipment, they cannot do much. On the other hand, the big and modern farm machineries/equipment, which are fuel saving and very economical, cannot be used in the Indian type of agricultural environment because they cannot be operated in a small individual plot of tiny area, which is unceasingly being fragmented into small sizes of landholding due to ever-growing pressure of population on the land. The average size of an agricultural holding in India is barely 2.6 hectares which is further segmented into numerous small plots. These very individual farming trends and conventions are a major cause due to which the agricultural yield of India are lowest in the world. On the other side, can it be believed that even less than 10 per cent of the American population engaged in agriculture not only produces enough for themselves and for their remaining 90 per cent but also has a huge surplus to feed nearly half of the world's populations. This is because America has adopted the intensive farming system under the corporate and organized sector (cooperative farming systems).

In India too, cooperative farming is the only possible solution by which Indian agriculture can be developed and revolutionized on more scientific lines and can be managed more efficiently. For adopting cooperative farming system and methods, India will have to first develop its unused and cultivable wasteland which is approximately 30 million hectares and will have to recognize agriculture as a priority sector industry by allotting the big farms to big industrial houses. The necessity to develop wasteland for the implementation of the aforesaid farming policy is that the Government cannot acquire the land already under the ownership of individual farmers, run through it is for more progressive purposes. Therefore, this is the only suitable way to develop largest tracts of waste and unused land and put it under organized farming systems, either public or private, in order to enable the country to export its agricultural surplus to meet the acute necessity of foreign exchange.

Equipment for Levelling Dotty Land Economically : Conventional Aspects/Modern Aspects

It is an established fact that large tracts of drastically uneven land cannot be levelled economically by conventional land levelling equipments like tractors, etc., because these levelling equipments can level very small areas of mildly uneven land. Besides this, the fuel consumption pattern per sq. meter is so high that it would not be wise to develop our huge area of wasteland by old levelling systems. The unsuitability of the prevailing levelling systems can be imagined simply by the fact that it takes minimum 6 to 8 minutes and 7 to 12 rupees to level merely a sq. metre of slightly uneven land.

The land in question, which is to be levelled, is situated between Mathura and Agra, having an area of 100 km × 100 km or 1,000,000 hectares approximately, and the most important is the natural gift endowed to the above piece of land it does not contain any stone or rocks, etc. It is highly fertile. The chemical analysis of the soil has revealed that the land can yield bumper crops of higher values in seven to eight years without the aid of any fertilizers. The entire area described above is full of dots which are very abruptly uneven, some dotts (a heap of soil raised conically upwards up to a certain height and having different dimensions) are even 8 to 9 metres high and somewhere 2-3 metres high, followed by 6 to 7 metres trenches. It can easily be worked out what the cost of levelling of above type of land, tremendously uneven, could be. With conventional levelling systems, it will not be an exaggeration if we say that such type of land cannot be levelled at all even if we totally ignore the cost and time factors described above.

Overhead Turning Leveller : A New Concept and a Single Solution

A modern method to develop such type of lands (uneven dots) by levelling is the "Overhead Turning Leveller" which levels highly uneven land from the top of the dot not containing stone, rocks, etc. The cost of levelling by the aforesaid system has been calculated at Rs. 2.50 per sq. mtr. or Rs. 25,000 per hectare, no matter how much abruptly uneven a land may be. Moreover, the three times levelling cost is included in the Rs. 25,000 per hectare rate because of the fact that after first levelling operation, levelled soil will subside during the rains and consequently land becomes again slightly uneven necessitating relevelling. The same happens during the next year (for one more subsequent rainy season), although to a least extent requiring third minor levelling operations.

A General Description of Overhead Turning Leveller

(a)	Operation	......	By 25 MW steam turbine
(b)	Type of wheels	......	Crawling chains of steel
(c)	Width of levelling bed	......	10 mtrs. wide
(d)	Crawling speed of entire system	......	15-20 km/hour
(e)	Overall length	......	40 mtrs.
(f)	Overall width	......	10 mtrs.
(g)	Overall weight	......	3,000 M.T.
(h)	Materials of construction	......	Mild steel mainly

Financial Aspects

A. Fixed Assets

	Name of Equipment	*Unit*	*Capacity*	*Price in Rs.*
(i)	Gas turbine complete with all accessories	1	25 MW or 30000 BHP	120000000
(ii)	Crawling chains of special sizes and quality			20000000
(iii)	Mild steel structurals like beams, angles, channels, bars, etc.of larger dimensions		515 M.T. @ Rs.9000 per M.T.	5000000
(iv)	Nine Nos. of high power tractors fitted with water tanks and six Nos. of high power tractors fitted with heavy duty trailers	15	@ 2,00,000 each.	3000000
(v)	Diesel jeeps with trolleys rear fitted for the purpose of site inspection and for bringing different spare parts/materials from near market	5	@ Rs.1,00,000 each	500000
(vi)	Ambassador cars	2	@ Rs.1,00,000 each	200000
(vii)	Motor-cycles	5	@ Rs.10,000 each	50000
(viii)	Salaries, fees and other expenses till two years, the leveller will be under fabrication, assembly, etc.			500000
(ix)	Fee for technical know-how from different			30000000

	field/branch of technologies, plus periodicals, literature, technical books, etc.	
(x)	Pre-operative and preliminary expenses of diverse kinds	20000000
(xi)	Miscellaneous fixed assets/infrastructures/ office equipments/ temporary dwellings for site workers, etc.	1250000
	Total	Rs. 250000000

Recurring Expenses (Working Capital) Per Month (25 days)

(A)	*Raw Material*	*Qty.*	*Rate Rs.*	*Price Rs.*	
(i)	Steam coal ash content not more than 20 to 25 per cent	6000MT	1350 MT	8000000	Aprox.
(ii)	Lubricant oil etc.	13MT	16,000 MT	200000	"
(iii)	Spare parts of levellers, all types	–	–	3000000	"
(iv)	Diesel oil including the cost of a small quantity of petrol at average rate of Rs. 4 a ltr.	2,00,000 ltres	Rs. 4 Per ltr.	800000	Approx.
(v)	Mobile oil, spare parts and maintenance of jeeps, cars, tractors and motor-cycles	–	–	200000	"
(vi)	Water for steam generation and its purification expenses, etc.	–	–	400000	"
(vii)	Miscellaneous stores and utilities, etc.	–	–	200000	"
	Total 'A'			12800000	"

(B) Labour and Supervision (per month) (25 working days)

	No.	Salaries per head Rs.	Total Rs.
(i) Honorarium of the Head of the Project	1	7500	7500
(ii) Mechanical Engineers (1 chief and 9 subordinates)	10	5000	50000
(iii) Electrical Engineers (1 chief and 4 subordinates)	5	5000	25000
(iv) Boiler and turbine Engineers and experts	30	7000	210000
(v) Fitters and helpers (mechanical turbine)	43	2000	86000
(vi) Fitters and helpers (electricals)	24	2040	40000
(vii) Fitters and helpers (boiler and turbine)	40	2540	100000
(viii) Semi skilled personnel (as required)	60	1400	60040
(ix) Unskilled personnel	150	800	120000
(x)G as cutlers and welders, etc.	20		200000
By adding 60% of the salaries towards E.S.I., P.F., Bonus, leave, etc.			431100
			Rs. 1149600
			or say Rs. 12,00,000

(C) Contingencies per month

(i) Travels and tours	Rs.	500000
(ii) Local conveyance	Rs.	100000
(iii) Technical study, consultation, experiments, books etc.	Rs.	300000
(iv) P & T expenses	Rs.	25000
(v) Miscellaneous	Rs.	75000
	Rs.	10,00,000

Total Working Capital per Month 'A' + 'B' + 'C'	=	Rs.	12800000 'A'
		Rs.	1200000 'B'
		Rs.	1000000 'C'
Total		Rs.	15000000

Area to be Levelled Per Day (Based on 20 hrs. per day operation)

In the general description of Overhead Turning Leveller on a previous page it has been said that the total width of levelled bed is 10 mtrs. This includes 2 mtrs. area levelled by scratcher arms and the crawling speed of the leveller is 15 to 20 km/hr.

Taking the crawling speed of the system, only 5 km/hr into account, the total area levelled in one hour is 5 (speed of the leveller in km) × width of levelling bed = 10 mtrs, therefore, 5 km. × 1000 mtrs. = 5,000 mtrs. × 10 mtrs. = 50,000 sq. mtrs. or 5 hectares, or in one day: 5 hectares × 20 hrs. = 100 hectares. In complete one year, there are minimum 200 such working days. Therefore, 200 days × 100 hectares = 20,000 hectares area levelled in one year.

Cost of Levelling

The proposed cost of levelling is calculated at Rs. 2.5 per sq. mtr. or Rs. 25,000 per hectare for three times levelling operation at only Rs. 1.50 out of Rs. 2.50 per sq. mtr. or at Rs. 15,000 per hectare out of Rs. 25,000 per hectare, leaving Re.1 per sq. mtr. or Rs.1,000 per hectare for subsequent (second year and third year) levelling operation, the cost of levelling of 2,000 hectares of land is Rs. 20,000 × Rs. 15,000 per ha = Rs. 30 crores.

Balancing the Total Expenses Incurred per Annum, Till the date of Completion of Levelling 20,000 hectares of Land and Proposed Cost of levelling at Rs. 15,000 per hect.

		Rs.	
(1)	Towards Raw Materials and Stores, i.e. Rs. 1,28,00,000 x 8 months.	102400000	The total return in terms of money earned by levelling 20,000 hectares of land at Rs.15,000 per hectare
(2)	Towards labour and supervision, i.e. Rs.12,00,000 per month x 8	9600000	
(3)	Towards contingencies, i.e. 10,00,000 per month x 8	8000000	
(4)	Towards accumulated interest on the investment on fixed assets, i.e. 25,00,00,000 at Rs.12% p.a. for 3 yrs. = Rs. 3 crores x 3 = Rs.9 crores	90000000	

(5) Towards interest on the working capital, i.e.12,00,00,000 at Rs. 15% per annum for one year. (Till 20,000 hectares of land is levelled)	18000000	= 20,000 x 15,000 = Rs.30,00,00,000
(6) Towards depreciation on the entire levelling system i.e. on 25,00,00,000@ Rs.15% p.a.	37500000	
(7) Towards provision for paying back the long-term borrowing (loan), i.e. 25,00,00,000 for fixed assests plus 12,00,00,000 for Working Capital = Rs. 37,00,00,000 to Indian exchequer, within 12 years, as a first instalment = 37 ÷ 12,00,00,000 = approx. Rs.3,45,00,000	34500000	
	Rs. 300000000	= 300000000

Returns in Terms of Agricultural Products sold, Produced in the 80,000 hectares of Levelled Land

A. Expenses for Producing/Growing 100 MT of Sugarcane and 20 MT of Potatoes of High Breed Varieties per hectare per Annum

Type of Expenses	*Exp. per hectare*
(a) (i) Tilling and cultivation	500
(ii) Soil dressing and purification by collecting weeds, etc.	100
	Rs. 600
(b) (i) Manures, nitrogenous fertilizers 250 kg. at Rs. 2 per kg.	500
(ii) Phosphatic fertilizers 200 kg. at Rs. 1.5 kg.	300
(iii) Potash 90 kg. at Re.1 per kg.	90
(iv) Other types of chemicals like growth regulators, rare minerals, salts, weedicides, insecticides, fungicides pesticides, etc.	150
	Rs.1040
(c) Seeds	
(i) Sugarcane, matured and healthy, 5 MT per hectare @ Rs. 200 per MT.	Rs.1000

(ii) Processing of fragmented sugarcane seeds, i.e. 5 MT@ Rs. 60 per MT including cutting the plant into small lengths of cobs, carrying it to the field and laying it by machine into the soil plus chemical treatment of seeds.	Rs. 300
(iii) Potato seeds of good quality and hybrid variety, culled properly, 1,500 kg. at Re.1 a kg	Rs.1500
(iv) Chemical processing of cut-potato and its plantation (putting it into soil)	Rs.300
(v) Irrigation, etc.	Rs.125
(vi) Managerial/administrative, travel, conveyance, P&T etc. stationery printing plus other miscellaneous expenses	Rs.75
	Rs.3300

Total 'A' = Rs.600 + total 'B' =1,040 plus Total 'C' = Rs.3,300 = Rs. 4,940 or say Rs. 5,000

Provisional Balance Sheet for First Year of Harvest and Sale of Sugarcane and Potatoes produced in the Levelled Land i.e. 20,000 hects.

The total return expected from 20,000 hectares of levelled land by producing sugarcane and potato is Rs. 64 crores.

The Rs. 64 crores is earned by growing sugarcane at the yield rate of 10 kg. per sq. mtr. or .01 MT per sq. mtr. or 100 MT per hectare and selling it at the rate of Rs. 170 per MT. Therefore, 20,000 hectares × 100 MT of sugarcane = 20,00,000 MT × Rs. 170 = Rs. 34 crores.

Potato produced at the yield rate of 2 kg. per sq. mtr. or 20 MT. per hectare and sold at Rs. 0.75 per kg., or Rs.750 per MT. Therefore, 20,000 hectares × 20 MT = 4,00,000 MT × Rs.750 = Rs. 30 crores.

The total sale of agricultural products produced in 20,000 hectares of levelled land is Rs. 34 crores by the sale of sugarcane plus Rs. 30 crores by the sale of potato. Total = Rs. 64 crores.

Total losses/expenses incurred till the date of harvest and the sale of 20 lakhs MT sugarcane and 4 lakhs MT potatoes produced in 20,000 hectares of levelled land.

(i)	Towards raw materials like coal, etc. for 8 months or 200 working days based year	Rs.120000000
(ii)	Towards accumulative interest on the investment for fixed assets, i.e. Rs. 25 crores at 12% p.a. for four years = 3 crores x 4 =12 crores	Rs. 120000000
(iii)	Towards accumulated interest on the working capital for raw materials and stores, etc., for levelling the said land, i.e. Rs.12 crores, at 15% per annum for two years = Rs.180 crores x 2 crores	Rs.36000000
(iv)	Towards accumulated interest on the working capital for cropping and plantation of 20,000 hectares of levelled land at the rate of Rs. 5,000 a hectare, i.e. Rs. 10 crores (interest at the rate of 15% per annum for complete one year, Rs. 150 crores)	Rs.15000000
(v)	Total expenditure incurred for planting 20,000 hectares of said land with sugarcane and potato, at Rs. 5,000 per hectare (including all types of expenses)	Rs.100000000
(vi)	Towards depreciation on entire leveller system the total cost of which is Rs. 25 crores, @ 15% per annum. for 2 years	Rs.75000000
(vii)	Miscellaneous losses	Rs. 5000000
		Rs. 471000000

Total Blocked Capital till the date of Harvest and Sale of Aforesaid Agricultural Products

(1)	In fixed assets (from last four years)	Rs. 250000000
(2)	In working capital during the course of levelling (in raw materials) from last two years.	Rs. 120000000
(3)	In working capital for plantation of sugarcane and potato in the levelled land, i.e. 20,000 hectares, for one year.	Rs. 100000000
(4)	Capital anticipated on other unseen and on unexpected heads	Rs. 30000000
		Rs. 500000000

Profit

Total return in terms of money earned by selling agricultural produce grown in aforesaid levelled land, Rs. 64 crores.	Rs. 640000000
Total losses/expenses till the date of income of Rs. 64 crores. = Rs.471000000	Rs.471000000
Profit before any type of taxes and instalment paid to be Financing Bodies/Agencies	Rs. 169000000

Completion Period of Project

The time required for design, fabrication, erection, assembly, trial operation of the overhead turning leveller is calculated two years approximately. After accomplishment of assembly and trial run of above system, it will take one year (based on 200 days, 25 days month, 20 hrs. a day, working system) to level 20,000 hecs. of land, and when the land is levelled, that is, ready for plantation, etc., one year will be taken by crops (sugarcane potatoes) to grow.

Phase-wise Programme for Completion of System–Overhead Turning Leveller Ready for Operation

First Phase

In the first phase, first three months, the site inspection and selection for setting a point of strategy, i.e. from where and from which direction the levelling should be started, will be determined. Further elaborate study of the physical stature of the reason mentioned in the Project Report will be conducted. The preliminary designing of the levelling system by consulting and discussing the concerned consultants will be prepared.

Second Phase

In the second three months, the more detailed technical calculations will be worked out. Final fabrication, erection and assembly drawing of the whole system will be prepared. The complete techno-economical study and appraisal of machinery/equipments, components and parts, etc., will be done. Complete/exhaustive technical specifications will be worked out for the purpose of advertising the requirements of machineries in concerned mass media to get quotations. At the end of this phase, the contract would be given to fabricators and erectors of the specific field after the appraisal and selection of their quotation/ proposal, techno-economically.

Third Phase

During the third three months, in the first week, the steel of required quality and specifications will be purchased and erection will be started at site. Meanwhile, technical discussions will be going on for the selection of 25 MW boiler and turbine plus its associated/requisite accessories, like pre-heaters, economizers, super heaters, heat recovery systems, and solar pre-heaters. Some of instrumentation systems will be discussed for the optimum result for controlling heat, pressure, etc. Instrumentation systems for the purpose of monitoring the programmes and progress will be selected.

Fourth Phase

During fourth three months, the selection will be made in order to equip with the overhead turning leveller with radio communication systems, digital control computers, modern instruments/meters/devices will also be selected after technical appraisal. To ensure a roof-plain surface, various types of instruments/indicators, meters will be proposed for import and the application for import licence will be submitted before the concerned authorities.

Fifth Phase

The turbine along with all aforesaid accessories will be brought to site for installation and at this stage, the sophisticated and complex erection and assemblies will be taken in hand. The installation of boiler, turbine, tanks, and other described instruments will be installed and assembled. These all activities will be continued till the end of the seventh phase, each phase comprising three months, thus reaching end of twenty-first month from the date of commencement.

Eighth Phase

During the last three months, i.e. after the end of first 21 months, a trial commissioning of the system will be undertaken and shortcomings and technical faults will be eliminated in order to ensure a smooth and unhampered operation.

Inauguration

During the last week of the 24th month, steps will be taken for the inauguration of the Scheme.

> *Note:* The rationale behind levelling the said land by spending such a huge amount of Money: One may ask as to why this particular land should be levelled first by spending such a huge capital and then it should be planted whereas there are millions of acres of naturally levelled land lying uncultivated and unused?

This is because the land almost like naturally levelled is not fit for cultivation due to acidic or alkaline composition of its soil. Therefore, such lands cannot be planted with sugarcane and potatoes or with any other commercial crops even after heavy investment in treating the land chemically. The heavy manuring and addition of chemical fertilizers are unable to make such lands suitable for cropping for 4-5 years. On the other hand, there is only one problem with the land being considered to be levelled that it is highly uneven.

As explained earlier in this Project Report, the aforesaid land is composed of highly fertile black soil having no stone or grit contents. The soil chemistry of the entire one million hectares of tract is very much favourable for foodgrain cultivation. Comparing this uneven land to other naturally even lands situated in different parts of the country, it is concluded that heavy yield can be harvested from the said land after developing it by levelling, which will augment the foodgrain production of the country and will enable us to export our surplus to earn huge amount of foreign exchange. But this is not possible in respect of other naturally levelled waste/useless land because heavy capital and long time will be required to develop the same and to take any benefit out of it.

ANNEXURE : 15.1

Explanations

1. Basis of Calculation for the Requirement of 6,000 MT of steam coal per month.
   1 MW= 1,000 KW or one MW/hour = 1,000 KW/hour
   1 MW/hr=3,500 BTU approx. or 1,000 KW/hour or 1 MW/hr. = 3,500 × 1,000 = 35 lakhs BTU of heat.

The capacity of the steam turbine (net output power) employed for the said overhead turning leveller is 25 MW or in one hour the requirement of heat in terms of BTU heat equivalent × 25 MW/hr. or 25 × 35 lakhs BTU = 8,75,00,000 BTU of heat is required in one hour of operation of OHTL system, or in 20 hours' operation, 8,75,00,000 BTU × 20 hrs. = 1,75,00,00,000 BTU of heat is required in one day based on 20 hours' operation.

∴ In one month, based on 25 working days in a month, the monthly BTU heat requirement = 1,75,00,00,000 × 25 BTU.

The heating value (calorific value of general bituminous coal, having not more than 15 to 18 per cent ash content, is roughly 13,000 BTU per pound of coal completely burnt in the presence of sufficient oxygen, or 1 kg = 22016 lbs. or BTU heat value of 1 kg of coal of above-mentioned quality = 2,2016 × 13,000 BTU = 28,610 BTU of heat per kg coal burnt. Assuming that different grades of coal will be supplied at different times, the average heating value should be 25,000 BTU per kg of coal. The total per month requirement of power (energy in terms of heat in BTU=

$$\frac{1{,}75{,}00{,}00{,}000 \times 25}{25{,}000 \times 1000}$$

=1750 MT of coal required per month.

Calculating the overall efficiency of boiler, steam turbine and other related equipments only at 30 per cent, the monthly requirement of coal comes to 100/30 × 1750 = 6,000 MT. approx.

Principle of Operation and Levelling

The main principle on which the said "Overhead Turning Leveller" will level the uneven land is that it will have two gigantic motorized arms at its two sides, fitted with teethed soil scratchers which, first will make levelled and plain path of one metre width at both the sides. In the arms, there will be heavy duty vibrators at both the sides for setting down the levelled soil strongly in order to prevent the down-soil-sinking of the heavy structure (the whole levelling system). The whole structure will then follow that very vibrated and levelled path crawling ahead for further levelling and so on. Uneven dots (area) left in between the two scratcher arms described above will be levelled by the main levelling system, i.e. main structure following the levelled and vibrated strong part being prepared simultaneously by scratcher arms. The main structure will level the remaining uneven area, i.e. of 8 metres width continuously. The levelling will be done in respect of 100 acre plots. Each plot will be made in such a way that it may be irrigated by a main water channel situated at a central point.

Note: The other details like drawings, basis of calculations for the requirement of 25 MW capacity turbine will be submitted later on. The break-up of expenses according to phase-wise programme will also be submitted after expert advice.

Some Unconsidered/Unaccounted Positive Factors in this Project

In calculating the return and profit at the end of the fourth year, several positive factors, sure to bring heavy dividend (income) in addition to figures already given in the Project Report, i.e. Rs. 16,90,00,000, are left unaccounted/overlooked in order to ensure absolute economical viability in the venture. For instance—

(1) The overhead turning leveller (DHTL) will completely level 100 hectares of land per day as discussed earlier in this Project Report, even at its 30 per cent working efficiency and capacity utilization and will continue levelling the land daily at the same rate. So this is clear that in the land, already levelled say during the first week, 700 hectares, we can go planting the sugarcane and potato or any other suitable short-life crop according to the order of the completion of levelling and season/climatic conditions, prevailing at the time. This will bring a small income as working capital for the time when Rs.10 crores will be required for planting 20,000 hectares of levelled land as explained before. Otherwise, the same sugarcane and potato can be used, if of hybrid variety, as a seed for the 20,000 hectares of land. This will substantially reduce the working capital requirement, i.e. Rs. 10 crores to almost one-fourth (Rs. 2.5 crores to Rs. 3 crores).

(2) Up to the time (till the end of the 4th year) the sugarcane and potatoes, as mentioned earlier, will be harvested from 20,000 hectares of levelled land, the levelling of the additional 20,000 hectares of land will already be accomplished which is worth Rs. 15 crores after deducting the cost of levelling, etc. This levelled land again will be put under the same crops or others thus bringing total income as a net profit. So all recurring expenses would be met from the first year of levelling and plantation.

(3) The green leaves and dried leaves of potato and sugarcane can be fermented into motor alcohol (ethanol) which is approxiinately 10 per cent of the total biomass, i.e. equal to 2 lakh metric tonnes @ Rs. 5,000 per MT which comes to Rs. 2,00,000 × 5,000 = Rs.1,00,00,00,000. All these profits are overlooked to remain at a safer side in this venture.

16

Fruits As a Component of Agroforestry

I.S. Singh and *R.K. Singh*

Fruits have special significance to human beings as a protective food. The well established and properly maintained orchard alone or grown with suitable tree intercrops can offer better yield and income per unit area as compared with field or cereal crops. Some fruits are hardy and are best suited to wastelands, arid and semi-arid regions for getting higher income with minimum inputs. Fruit plants also provide aesthetic beauty, purify air and decrease pollution. Present chapter discusses the uses and suggests suitable fruit crops and cropping models for economic utilization of wastelands for socio-economic development of the rural people.

Key Words : Wastelands, fruit forestry system, productivity, fruit crops, energy value

Importance of fruit in human diet is well recognised. As compared to many developed countries, and even some developing countries, our per capita per day consumption of fruit is one of the lowest (45 gm), being much below the minimum dietary requirement (85 gm) Malnutrition being prevalent in the rural areas, the fruits, therefore, have special value. Fruits also contribute to the aesthetic side or rural and home life of the community. Plantation of trees which are useful as fuel, fodder and timber is very much emphasised under agro or social forestry programme. However, little or no attention is being paid for fruit trees.

One must note that the economic age of forest species is longer than the fruit trees. The small or marginal farmers would like to grow something which can support their family in short time. Thus, fruits appear to have better prospects in agroforestry system in terms of health and economy as compared to cereals and forest species. Present chapter discusses the needs that fruit should get priority in prevailing concept of growing trees for fuel, fodder and timber under agroforestry or plantation programme.

Protective Food

Foods rich in vitamins and minerals are termed as protective foods. The fruits are the chief source of vitamins which help in the maintenance of proper health and resistance to diseases. Fruits are also good source of minerals like calcium, iron, etc., the deficiency of which may lead to disturbances in the metabolism and can cause several ailments. Most of cereal grains are poor in vitamins (Table 16.1) and minerals (Table 16.2).

Table 16.1 : Vitamins content of some cereals and fruits

Cereals/Fruits	*Vitamins*			
	A (IU/100g)	*B_1 (mg/100g)*	*B_2 (mg/100g)*	*C (mg/100g)*
Rice	0	0.06	0.06	0
Wheat	49	0.45	0.17	0
Mango	5000	40.00	50.00	45
Papaya	2100	40.00	250.00	50
Aonla	160	30.00	0.10	650

Table 16.2 : Mineral content of some cereals and fruits

Cereals/Fruits	*Minerals (mg/100 g)*	
	Iron	*Calcium*
Rice	3.1	10
Wheat	4.9	41
Karonda	39.1	160
Phalsa	3.1	129

Higher Yield and Energy Value

The well established and maintained fruit plantation can offer better yield and energy value per unit area as compared to field crops. Fruits such as aonla, guava, papaya, etc. can provide higher yield and energy value per hectare as compared to cereals like wheat and rice (Table 16.3).

Table 16.3 : Comparative yield and energy value of some cereals and fruit crops

Crops	*Yield (t/ha)*	*Energy value (K cal/ha)*
Rice	1.35	4657500
Wheat	1.80	5623090
Aonla	20.10	11600000
Guava	8.70	5748600
Papaya	50.00	15000000

Early Yield

The budded fruit plants start bearing and giving yield at an early age. Data recorded on the average yields of 2 year old plantation of guava, ber, grape, papaya and phalsa are presented in Table 16.4. The early yields of these plants are very encouraging, which can be a good source of income to the farmers at very early stage. The same is not possible with any forest species.

Table 16.4 : Productivity of 2 year old plantation of some fruit crops

Fruit	*Productivity*	
	Per plant (kg)	*Per hectare (Qtls)*
Guava	10.0	26.0
Ber	8.0	12.0
Grape	5.0	50.0
Papaya	30.0	600.0
Phalsa	0.5	10.0

Fuel Wood

Some fruit crops require annual pruning for their balanced growth and fruiting. Data recorded on dry wood of some of the fruit crops after pruning is furnished in Table 16.5. Phalsa and mulberry produced enough fuel wood at an early age which can be an additional source of income to the farmers. There is burning problem of fuel in the villages and regeneration characteristics of these fruit plants suggest the possibility for using these plants in agroforestry system.

Table 16.5 : Fuel wood (pruned material) of some fruit crops

Fruit crop	*Age (year)*	*Dry wood*	
		Per plant (kg)	*Per hectere (Qtls)*
Phalsa	2	2.0	40.0
Ber	2	5.0	7.5
Mulberry	4	5.0	31.2

Suitable Means to Utilize Wastelands

Wastelands are available in plenty in India. Some fruits can offer best utilization of wastelands for getting higher income with minimum inputs. Crops like aonla, ber, guava, karonda, phalsa, jamun, grape, bael, etc. can be grown in such area. Data obtained on salt tolerance and productivity of these fruits are presented in Table 16.6. Productivity of these fruit trees seems to be high even in problematic soils. Therefore, such fruit plants can be ideally suited for agroforestry models to utilize wastelands which are available in villages. This will go a long way in enhancing the wealth and prosperity of the farmers.

Table 16.6 : Salt-tolerance and productivity of some fruit plants at full bearing age

Fruit crops	*Salt tolerance (phe)*	*Age (years)*	*Productivity (t/ha)*
Aonla	10.0	8 and above	20.5
Karonda	10.0	5 and above	5.2
Ber	9.5	5 and above	15.5
Jamun	9.5	6 and above	16.0
Phalsa	9.2	3 and above	6.0
Grape	9.0	4 and above	18.3
Guava	9.0	4 and above	12.5
Bael	8.5	8 and above	16.5

Suitable for Fruit-Forestry System

Looking into the needs of the farmers, some fodder or fuel species such as *Eucalyptus, Casuarina* and *Leucaena* can be grown between the rows of aonla, jamun, ber and guava to utilise the interspaces. These plants can be taken out by the time fruit plants reach full bearing. *Casuarina* is very hardy and nitrogen fixing plant which can be grown in salt-affected soil to improve the fertility of the orchard.

Better Employment Opportunity

The major factor contributing to poverty in the rural area is the unemployment which particularly effect the poorest segment of the rural population. Hence, the most urgent need in rural India today is the diversification of opportunities for employment and income. Plantation of fruit crops under agroforestry programme can provide considerable employment opportunities both during the production and post-harvest phases.

17

Prospects of Aonla, Ber and Bael Cultivation on Salt-Affected Wastelands in India

R.K. Pathak, S.N. Dikshit and *R.Dwivedi*

Aonla, ber and bael are hardy fruit crops which can successfully be cultivated in the problematic soils having predomination of various salts. Studies conducted at NDUAT clearly indicate that seed germination of these species on salt-affected soil is poor, hence, its sowing should be avoided. On the other hand, six months to one year old seedlings and budded plants survive satisfactorily. Aonla and ber plants can tolerate 33 and 45 per cent exchangeable sodium content or 12.8 and 15 mmhos/cm ECe respectively, while the tolerance limit of bael has been found up to 30 ESP. Beyond this, incorporation of soil amendments and leaching of soluble salts will be helpful in improving the soil conditions. Promising varieties/strains of these fruit crops have been selected and agro techniques, *viz.*, size of pits, soil mixture and vegetative propagational methods have been standardized. It is suggested that in mass plantation programme, instead of monoculture of forest species, 15-20 per cent fruit plants must be included, particularly on the salt-affected soils and other wastelands.

Key Words: Seed germination, sodic soil, saline soil, budded plants, ESP, ECe, GR.

Salinity and alkalinity/sodicity are the major problems in arid and semi-arid regions of the country, more prominently in the Indus and Gangetic plains in the north. Out of the total salt-affected area of about

7.0 million ha in India, 1.3 million ha alone is spread in Uttar Pradesh in the form of wasteland as well as in small patches within the cultivated areas (Agarwal *et al.*, 1979). Today, much emphasis is being given on the management of wastelands after establishment of the National Wasteland Development Board by the Central Government. But priority is given for the plantation of forest trees. However, there is every possibility of utilizing these barren areas by fruit cultivation along with forest trees.

Fruits have special significance to human beings as a protective food. Their yield and income per unit area are much higher as compared to cereals and many forest trees. In view of these facts, studies on salt tolerance of aonla, ber and bael have been conducted at this University which clearly indicate that these fruit crops can successfully be cultivated on salt-affected soils.

Seed Germination

Studies on seed germination of aonla and bael were conducted at different ESP (Exchangeable Sodium Percentage) levels of sodic soils maintained with $NaHCO_3$ in normal soil (Table 17.1). Further, seed germination studies were conducted on ber and bael in reclaimed sodic soil with gypsum as per GR (Gypsum Requirement) values (Table 17.2).

Table 17.1 : Effect of sodicity on seed germination and seedling survival in aonla and bael

Aonla			*Bael*		
ESP	*Seed germ. (%)*	*Seedling survival (%)*	*ESP*	*Seed germ. (%)*	*Seedling survival (%)*
Control	38.7	34.2	Control	96.2	96.2
14.3	25.0	19.2	14.8	41.2	25.0
32.7	19.7	8.7	29.0	0.0	0.0
47.3	6.7	0.0	43.2	0.0	0.0
64.4	4.2	0.0	58.9	0.0	0.0
77.7	0.0	0.0	–	–	–

Results of these studies indicated that seed germination of aonla (Tewari, 1983), ber (Pandey, 1984) and bael (Tewari, 1985) in sodic soil is not practicable because of poor germination. Hence, seed sowing should be avoided in salt-affected soils.

Table 17.2 : Effect of sodicity on seed germination of ber and bael

G R levels	*Germination percentage*			
	Ber		*Bael*	
	Mean	*Angl. values*	*Mean*	*Angl. values*
Control	2.7	9.3	3.0	10.0
25% GR	15.1	22.9	12.1	20.3
50% GR	36.0	36.9	30.5	33.5
75% GR	45.7	42.5	43.2	41.1
100% GR	59.6	50.6	56.3	48.6
S. E. ±	–	1.0	–	1.7
C. D. at 5%	–	2.2	–	3.8

Establishment/Survival of Seedlings and Budded Plants

Studies were undertaken on seedlings/budded plants in sodic and saline soils. In aonla and ber, seedlings as well as budded plants of cultivar Chakaiya and Banarasi Karaka were evaluated at different ESP and ECe (electrical conductivity of saturation extract) levels, respectively.

Results indicated that aonla seedlings as well as cultivar Chakaiya showed a good survival up to 32.9 ESP and 10.2 mmhos/cm ECe level (Tables 17.3 and 17.4). Since, at the culmination of experiment, the salinity level was found to be increased from 10.2 to 12.8 mmhos/cm with a survival of 87.5 per cent of seedlings and 75 per cent of budded plants, the tolerance limit of aonla has been fixed up to 12.8 mmhos/cm ECe level (Dikshit, 1987).

The performance of seedlings as well as of budded plants clearly indicated that ber can successfully be cultivated in sodic soil up to 45 ESP and in saline soils up to 15 mmhos/cm ECe level (Tables 17.5 and 17.6).

In bael, salt tolerance studies were conducted in two types of sodic soil. First, seedlings were grown at different ESP levels of sodic soil artificially maintained with $NaHCO_3$ in the normal soil. Further, the study was repeated in reclaimed sodic soil with gypsum as per GR values.

Results (Table 17.7) revealed that bael can successfully be grown in sodic soils up to 30 ESP (Tewari, 1985) and 50 per cent GR levels (Pandey *et al.*, 1985). However, there is urgent need for varietal evaluation of aonla, ber and bael in sodic and saline soils so that limit of salt tolerance may be assessed.

Table 17.3 : Effect of ESP levels on per cent plant establishment and survival of aonla

	Seedlings		*Cv. Chakaiya*	
ESP levels	*Establishment (%)*	*Survival (%)*	*Establishment (%)*	*Survival (%)*
Control	100.0	100.0	100.0	100.0
14.7	100.0	100.0	100.0	100.0
32.9	100.0	87.5	100.0	75.0
46.5	87.5	62.5	5.0	50.0
S. E. ±	–	11.9	–	10.1
C. D. at 5%	NS	26.0	NS	22.0

Table 17.4 : Effect of ECe levels on per cent plant establishment and survival of aonla

	Seedlings			*Cv. Chakaiya*		
ECe levels (mmhos/cm)	*Estab. (%)*	*Surv. (%)*	*Transf. value*	*Estab. (%)*	*Surv. (%)*	*Transf. value*
Control	100.0	100.0	10.0	100.0	100.0	10.0
5.4	100.0	100.0	10.0	100.0	100.0	10.0
10.2	100.0	87.5	9.3	87.5	75.0	8.5
15.5	62.5	0.0	1.0	50.0	0.0	1.0
S. R. ±	13.7	–	0.3	8.3	–	0.5
C. D. at 5%	30.0	–	0.7	18.0	–	1.0

Table 17.5 : Effect of ESP levels on per cent plant establishment and survival of ber

	Seedlings		*Cv. Banarasi Karaka*	
ESP levels	*Estab. (%)*	*Surv. (%)*	*Estab. (%)*	*Surv. (%)*
Control	100.0	100.0	100.0	100.0
15	100.0	100.0	100.0	100.0
30	100.0	100.0	100.0	87.5
45	100.0	87.5	87.5	75.0

Table 17.6 : Effect of ECe levels on per cent plant establishment and survival of ber

ECe levels (mmhos/cm)	*Seedlings*		*Cv. Banarasi Karaka*	
	Estab. (%)	*Surv. (%)*	*Estab. (%)*	*Surv. (%)*
Control	400.0	100.0	100.0	100.0
5	100.0	100.0	100.0	100.0
10	100.0	100.0	100.0	87.5
15	87.5	75.0	75.0	62. 5

Table 17.7 : Effect of sodicity on seedling survival of bael

ESP levels	*Survival (%)*	*Angl. value*	*GR levels*	*Survival (%)*	*Angl.* value
Control	93.7	82.5	Control	62.5	52.5
14.8	93.7	82.5	25%GR	63.7	56.2
29.0	81.2	71.2	50%GR	81.2	71.2
43.2	0.0	1.8	75% GR	100.0	90.0
58.9	0.0	1.8	100%GR	100.0	90.0
S. E. ±	–	9.7	S. E. ±	–	8.2
C. D. at 5%	–	20.8	C.D.at 5%	–	18.0

Agro-Techinques

Besides the tolerance limit, attempts have also been made to standardize certain agro-techniques which help in establishment and growth of the plants under sodic/saline soil situations.

Cultivation in highly sodic soil requires reclamation of soil by incorporation of gypsum/pyrite as per GR values and in saline soil through leaching of soluble salts. Soil amendments should be applied in the pits only. The size of pit depends upon the type of soil and plant. The sodic/saline soils which are characterized by the presence of indurated clay and *kankar pan* ($CaCO_3$), breaking of *kankar pan* is must before filling of the pits. The soil mixture consists of top or upper soil, sand, FYM/compost and aldrin against termite which is serious problem on wasteland. Pits should be dug in May/June and rainy water should be allowed to collect in the pits. Thus, collected dirty water contains a lot of soluble salts which must be thrown outside. The details of pit preparation and planting are given in Table 17.8.

Table 17.8 : Details of pit preparation and planting of aonla, ber and bael

			Soil-mixture				
Fruit crops	*Cultivars*	*Pit size (cubic m)*	*Soil*	*F.Y.M. (kg)*	*Sand (kg)*	*Gypsum/pyrite (kg)*	
Aonla	Chakaiya Kanchan Krishna	1.00	Top soil	40	30	5	4
Ber	Gola Banaras-Karaka Umran	0.75	Top soil	30	20	5	4
Bael	Faizabad Selection	0.75	Top soil	30	20	5	4

Sometimes, budded plants may pose a problem in their initial establishment. Hence, six months to one year old seedlings may be transplanted in the prepared pits and after their establishment, these seedlings must be top worked through budding with promising varieties of aonla, ber or bael.

Precautions

Besides the above-mentioned agro-techniques, following points should also be taken into consideration for proper utilization of salt-affected soils for fruit cultivation.

(i) Closer plantation

It is a matter of common experience that there is poor growth in salt-affected soils as compared to the normal soil. Hence, planting distance should be reduced as compared to normal soil conditions.

(ii) Soil covering with inter/cover crops/mulching

Closer plantation coupled with inter-cultivation of inter/cover crops or mulching shall minimize the evaporation and transpiration losses. These shall also be helpful in improving the organic matter content, keeping weed population under control, improving the soil fertility, growth and productivity of the plants. Cultivation and turning of green manure crops, *viz.*, dhaincha (Sesbania aculeata) is also very much helpful in improving the soil conditions.

(iii) Increased frequency of irrigation

Because of poor built-up of soil, the water holding capacity of such soil is very poor, prolong irrigation will result in upper movement of salts. Hence, light irrigation is advisable at less intervals. Irrigation should be done with sweet water. Recent irrigation technique, *viz.*, drip irrigation may prove more beneficial than conventional methods.

(iv) Careful selection of fertilizers

Fertilizers having acidic reactions, *viz.*, ammonium sulphate, urea, calcium ammonium nitrate, super phosphate and potassium sulphate should be preferred. Zinc deficiency is very common in salt-affected soils. Hence, $ZnSO_4$ should either be sprayed or mixed along with other fertilizers.

REFERENCES

Agarwal, R. R., Yadav, J. S. P. and Gupta, R. N. (1979). *Saline and alkali soils of India.* I.C.A.R., New Delhi, p.11.

Dikshit, S, N, (1987). *Salt tolerance studies in aonla (Emblica officinalis Gaertn).* Ph. D. Thesis, N. D. Univ, of Agric. & Tech., Faizabad.

Pandey, D. (1984). *Effect of sodicity on germination, leaf injury, growth and nutrient uptake of ber (Zizyphus* spp.*).* M. Sc. (Ag.) Thesis, N. D. Univ. of Agric. & Tech., Faizabad.

Pandey, G., Pathak, R. K. and Om, Hari (1985). "Effect of sodicity on mineral composition, chlorophyll content and injury symptoms in bael *(Aegle marmelos Correa)* leaves." *Prog. Hort.* 17 (1) : 5-8.

Tewari, R. K. (1983). *Studies on seed germination and seedling growth of aonla (Emblica officinalis Gaertn) in sodic soil.* M. Sc. (Ag.) Thesis, N. D. Univ. of Agric. & Tech., Faizabad.

Tewari, R. N. (1985). *Efect of ESP levels on germination, leaf injury, growth and nutrient uptake of bael (Aegle marmelos Correa).* M. Sc. (Ag.) Thesis, N. D. Univ. of Agric. & Tech., Faizabad.

18

Experiences of Growing Crops and Trees on Salt-Affected Soils of Hardoi District in Uttar Pradesh

K. K. Mehta

Reclamation of salt-affected soils of Alisabad and Sikanderpur villages of Hardoi district was undertaken during 1986. The soil analysis revealed that soil with pH of 10.5 and ECe of 24.0 dSm^{-1} was highly saline and alkali in nature at the surface. The water-table fluctuated between 60 cms to 150 cms. The technology evolved at C.S.S.R.I., Karnal was adopted for the reclamation of soil and crop cultivations. It was observed that during the first year of reclamation, an average yield of 30 q/ha of rice (unhusked) and 8 q/ha of wheat was obtained. About 20,000 tree saplings mainly of *Prosopis juliflora*, Ceres, Jangal-jalebi and Kanji were planted after the treatment of soil of pits (60×60×60 cms) with 10 kg sand and 5 kg pyrites. Among these, *P. juliflora* was found more tolerant with regard to survival and growth. Keeping in view the improvement in soil properties, it is proposed to revise the gypsum requirement recommendation under high water-table conditions.

Key Words : Usar lands, reclamation of soil, *Prosopis juliflora*, water-table.

In Uttar Pradesh, salt-affected soils which are locally called 'Usar soils' are spread over 12 lakh hectares of land (Abrol and Bhumbla, 1971). These soils contain high quantities of salts mainly sodium carbonate due to which they have very high pH, high exchangeable sodium, dispersed soil, low infiltration rate, low organic carbon, low available nitrogen and very less soluble calcium.

In the present chapter, performance of crops and trees and economics of soil reclamation and crop cultivation on the salt-affected soils of Alisabad and Sikanderpur villages in Hardoi district are examined. The Panchayat alkali lands in these villages were given to the poor farmers on 'patta' basis (ownership rights) and all the operations starting from the land preparation to crop harvesting were done by the farmers after getting training at C.S.S.R.I., Karnal.

Characteristics of the Salt-Affected Soils

Analysis of soil samples carried out by Singh *et al.* (1986) Bhargava and Abrol (1978) and at soil testing laboratory, Hardoi revealed that the soil with a very high pH (> 10) and salt content (> 20 mmhos/cm) at the surface was highly deteriorated. At many places, the salt layer was thick and loose. There was hard 'kankar' layer at about 80 cm depth. The calcium carbonate content was 4 per cent at the surface and 16 per cent at one metre depth. There was deficiency of organic carbon and available nitrogen.

Water-table measured in May, 1986 from the open wells was at 1.5 m at 500 metres distance from the Sharda canal and at 2.4 m at 1500 metres away from the canal. However, during rains water-table used to rise near the soil surface. The quality of underground water is good having electrical conductivity of 600 micromhos/cm.

Reclamation of Soils

The technology evolved at C.S.S.R.I., Karnal with little modifications was used for the reclamation of soil :

1. Installation of tubewells to lower down the water-table and arrange assured irrigation. A total of 11 bores were made under free boring scheme. Two of the farmers got loan for the purchase of diesel engine at subsidised rates. All the

tubewells were diesel operated as the farmers preferred them due to irregular supply of electricity.

2. As the irrigation on undulating fields without strong bunds was not only difficult but also expensive, fields were bunded and levelled. Most of the land levelling was done only after rains in the month of June.
3. A surface drain of about 800 meter length was dug by the Sarvodaya Ashram and villagers on voluntary basis so that in case of heavy rains, excessive water could be taken to the nearby 'Bhainshta drain'.
4. After levelling and ploughing, gypsum powder @12.5 t/ha was applied by broadcasting. It was mixed in the upper 10 cms soil and irrigation was given to keep water standing.
5. Rice nursery of variety Jaya was grown at one place for all the farmers. When it was one month old, transplanting was started from 1st July, 1986. About 3-4 seedlings per spot/hill were planted and distance from hill to hill was kept about 15 cms.
6. A total of 38 kg N and 25 kg zinc sulphate per ha were applied as basal dose. Nitrogen was applied in the form of urea. No P or K was applied as these soils contained sufficient quantities of these nutrients. A total of 75 kg N and 15 kg zinc sulphate per ha were applied at tillering stage and 37 kg N/ha was applied at penicle initiation stage.
7. Insecticides and fungicides were also applied to check the attack of insects and bacterial blight.
8. Crop was harvested in the month of October. Harvesting and threshing were carried out by the farmers themselves.
9. During winter, wheat HD-2009 and HD-1553 were sown.
10. During summer (i.e. last week of April to first week of May), Dhaincha (Sesbania aculeata) was grown for green manuring.
11. During 1987, rice was transplanted again without further addition of gypsum.

PERFORMANCE OF DIFFERENT CROPS

Paddy and Wheat Yields During First Year

The average grain yield of paddy was 30 q/ha with a range of 10 to 50 q/ha. The yields were equally good in comparison to the average yields of paddy in the surrounding area. The crop yields in many plots was good due to the fact that farmers got the lands levelled and did most of the farm operations in time.

In case of wheat crop, the average yields were 8 q/ha. The yield was, however, lower than the yields obtained in Haryana. The reason for this poor yields in few cases was that the initial alkali and salinity levels were very high and thus one year was not sufficient to improve the soil for the wheat yield. The soil was also heavier in texture and the water-table was high. In addition to these factors, land levelling could not be done to the desired level as there was very less time available for this purpose after first rainfall.

Experiences of Growing Dhaincha

The seeds of dhaincha germinated well and the plants grew very well. However, as the season was dry, the crop required about 6 irrigations with a total cost of Rs. 1259 per ha. In addition, an expenditure of Rs. 850 per ha was incurred on the purchase of seeds, sowing, fertilizer and burying of dhaincha.

Under such conditions where irrigation is applied through diesel operated tubewells, it is suggested that after harvesting wheat crop, the fields be ploughed and left as such. This ploughing will help in breaking the soil capillaries which in turn will check the upward movement of water and salts.

Alternatively sorghum (jowar) can be grown for fodder after harvesting wheat crop. This crop can be harvested by mid-July and then Basmati rice (B-370) can be transplanted. This crop rotation has also been successfully adopted by many farmers in ORP villages near Karnal in Haryana (Mehta, 1983).

Paddy Yields During Second Year

The average yield of paddy in 1987 was 27 q/ha which was less by 3 q/ha as compared to the average yield of previous year. The decrease in yield was mainly due to severe drought conditions.

Improvement in Soil Properties

pH and salt concentration in the soil decreased after one year of reclamation (Table 18.1). As the original pH and salt concentration was very high, normal crop yields can be obtained by growing rice and other crops in the subsequent years.

Table 18.1 : Changes in soil properties due to reclamation in Alisabad

Soil depth (cm)	*Before reclamation (July, 86)*		*After reclamation (Feb., 87)*	
	pH (1:2)	*ECe (mmhos/cm)*	*pH (1:2)*	*ECe mmhos/cm)*
0-15	10.5	24.8	9.6	4.8
15-30	10.4	14.4	10.0	7.2
30-45	10.4	10.0	10.2	8.0

As a result of using the tubewells for irrigation, the water-table was lowered after one year of reclamation (Table 18.2). These cavity wells served as vertical drains. During reclamation water-table should be lower beyond a critical depth so that leached salts do not come up to the surface again. For these areas the critical depth is 3 m approximately.

Table 18.2: Changes in water-table at Alisabad and Sikanderpur villages

	Depth (m)	
	Alisabad	*Sikanderpur*
May, 1986	1.5	2.4
May, 1987	2.4	3.6

It was also observed that running of 10 cm diameter tubewell for 7-8 hours lowered the level of water by 30 cm in an open well at a distance of 30 metres. However, after stopping the tubewell, the water level comes to its original depth.

Economics of Soil Reclamation and Crop Cultivation

Economics of reclamation was calculated from 15 randomly selected farmers including 13 farmers who did not install tubewells and 2 farmers who purchased their own diesel engines for the tubewells. During first year of reclamation, all the inputs were provided to the farmers free of cost. This was done by making available the various subsidies on tubewell installation, land levelling and gypsum application. The remaining cost was arranged by S.P.W.D. (Society for Promotion of Wastelands, New Delhi). The cost incurred on different inputs and farm operations have been presented in Table 18.3. Tubewell bores were made free of cost under 'Free Boring Scheme' which was available for the Scheduled Caste farmers. Two farmers were presuaded to purchase their own diesel engines for operating the tubewells by taking loan from the bank. There was subsidy up to 25 per cent on diesel engine. Those farmers who owned tubewells sold water @ Rs. 12 per hour to others. However, this amount was paid to them by S.P.W.D. for the first year. Gypsum was available @ Rs. 140 per tonne on 75 per cent subsidy. On an average gypsum worth Rs. 1750 per ha was applied. This cost was also borne by S.P.W.D.

An initial investment of Rs. 15,100 per ha was needed provided various resources as well as different operations are available on subsidized rates. However, in case no subsidy was provided to the farmer this initial investments may increase by one and half times.

Cost of rice and wheat cultivation was Rs. 3900 and Rs. 2950 per ha, respectively when the farmers had their own tubewells. However, the cost was Rs. 5100 for rice and Rs. 3450 per ha for wheat when the water was hired from the nearby farmer.

The total return from rice and wheat were Rs. 4500 and Rs. 1280 per ha, respectively during the first year. The return from wheat crop was less which may be due to the fact that some of the fields could not be levelled properly. In addition, the initial pH and salinity levels were also high. Even after one year, the EC during wheat crop was 4.8 mmhos/cm and pH was 9.6 which were not favourable for wheat crop.

Trees and Grass Plantation

About 10 ha of salt-affected land of Alisabad and Sikanderpur villages were put under tree plantation by the Forest Department. This land existed in five blocks. Each block was bunded by digging one meter

Table 18. 3 : Cost of soil reclamation and crop cultivation during first year

Particulars		*Cost per ha (Rs.)*
Soil reclamation		
Bunding		250.00
Levelling		2000.00
Tubewell boring and accessories		4000.00
Diesel engine (8 H.P.) on subsidized rates		6500.00
Tubewell room		Not constructed yet
Gypsum cost @ 12.5 t/ha at subsidized rate		1750.00
Gypsum transportation, application & mixing		600.00
	Total	15100.00
Rice cultivation		
Nursery raising		500.00
Seedling uprooting, transplanting		350.00
Irrigations (own tubewell). However it was Rs. 2000/ha if purchased from nearby tubewell		800.00
Fertilizers		1000.00
Insecticides/fungicides		250.00
Labour cost		500.00
Harvesting and threshing		500.00
	Total	3900.00
Wheat cultivation		
Field preparation		400.00
Seed and Sowing		500.00
Fertilizers		1000.00
Irrigation (own tubewell)		400.00
Harvesting and threshing		250.00
Labour cost		400.00
	Total	2950.00

wide trench outside the bund. The trench was made to check the entry of animals. During May-June 1986, pits of 60 × 60 × 60 cms were dug. After few days, the pits were refilled with soil by treating with 5 kg pyrite and 10 kg sand. Four tree species, viz. *Prosopis juliflora,* Ceres, Kanji and Jangaljalebi were planted. It was observed that among these four species, *P. juliflora* was the most tolerant and its survival percentage was maximum. Due to high salinity and alkalinity and low rainfall, the mortality was high during the initial 2-3 months and thus about 10,000 plants were replanted in the area. In February, 1987 about 10,000 plants were again replaced. The replacement was done mainly with *P. juliflora.* Except in the severely affected areas, where the pH was around 10.4 and ECe more than 15 mmhos/cm, the plants have good survival. They are growing very well in many patches and after some of the individuals have gained girth in between 6 cms to 15 cms with many side branches.

Para grass and Napier grass were planted on a small piece of salt-affected land after treating the field with gypsum. These grasses grew well only in fields or channels where standing water could be maintained. During *Kharif* 1987, para grass cuttings were also planted in between the *P. juliflora* plants. It was irrigated but it had to pick up growth yet. Before reaching to any conclusion it has to be seen how economic it will be by keeping into consideration the large number of irrigations needed for this species.

Acknowledgements

The author is grateful to Dr. R. C. Mondal, Director, C.S.S.R.I., Karnal for encouragement and guidance during these studies and to Dr. G. P. Bhargava, Head, Division of Soils and Agronomy and Dr. A. K. Agnihotri, Scientist S-1, Economics for their help during the preparation of this manuscript. Thanks are also due to S.P.W.D. and Voluntary Agency Staff for providing necessary help in carrying out the studies.

REFERENCES

Abrol, I. P. and Bhumbla, D. R. (1971). "Saline and alkali soils in India. Their occurrence and management." *World Soil Resource, F. A. O. Report,* No. 41 : 42-51.

Bhargava, G. P. and Abrol, I. P. (1978). "Characteristics of some typical salt-affected soils of Uttar Pradesh." *CSSRI, Report No.* 6, p.31.

Mehta, K. K. (1983). *Reclamation of alkali soils in India.* Oxford and IBH Publication Co., New Delhi, p. 283.

Singh, D. R , Das, S. N. and Karale, R. L. (1986). *Report on detailed soil survey of 'Usar' lands, Hardoi District (U. P.).* Report of All India Soil and Land Use Survey Organisation, Deptt. of Agri. and Rural Development, IARI, New Delhi, p. 10.

19

Effect of Mango (*Mangifera Indica* L.) Trees on Growth and Yield of Wheat (*Triticum Aestivum* L.)

A. K. Saxena, P. K. Singh and *B.P. Singh*

Five individuals of mango (*Mangifera indica L.* var. Dashehari) from each of the 5 sites were selected, and these were measured for tree height, canopy depth, canopy width and stem diameter. The wheat (*Triticum aestivum* L. var: HD 2285) under the canopies of mango trees and in adjacent open areas was sampled for certain growth parameters and yield of grains and by-products. Light intensity was also measured under the tree canopies and in open areas at the time of crop harvesting. The average height of mango trees ranged from 2.2 m on the site with 6 year old trees to 11.2 m on site with 25 year old trees. Relative light intensity under tree canopies decreased with increasing canopy depth and tree height. There was a marked reduction in growth and yield of wheat under the tree canopy. Wheat plants indicated 5.7 per cent decrease in total population, 23.5 per cent in tiller number, 4.6 per cent in height, 11.3 per cent in panicle length and 21.6 per cent in number of grains per panicle. The reduction in grain yield under tree canopy varied from 14.5 per cent on the site with 6 year old mango trees to 60.7 per cent on site with 25 year old mango trees. The yield shows a positive relation with light intensity. Average grain yield which varied between 5.9 q/ha and 16.5 q/ha among various sites indicates a negative relation with the tree canopy cover.

Key Words: Grain yield, light intensity, *Mangifera indica*, *Triticum aestivum*, under canopy.

Mango (*Mangifera indica* L.) is the most characteristic species of Tropical semi-evergreen and Northern secondary moist mixed deciduous forests of Orissa (Champion and Seth, 1968). This species is also very well adapted to tropical and sub-tropical climates and it thrives well in almost all regions from sea level an altitude of about 600 m. Mango occupies about 42.6 per cent of the total area under fruit species in India (Anonymous, 1983). Among the various states, Uttar Pradesh has the largest area under mango (33.2%) followed by Bihar (13.6%) and Andhra Pradesh (13.5%).

Since mango trees take several years to grow to their full size, intercropping in order to utilize the interspaces is very important. Generally, intercrops such as wheat and rice are grown during the early stages of mango plantation. However, no systematic study was made earlier to examine the effect of mango trees on the associated agricultural crops. The present study deals with the effect of mango trees on the growth and yield of wheat.

Study Area and Methods

The study was carried out at farmer's fields situated at Mau Shivala, about 5 km from Faizabad in eastern part of Uttar Pradesh. The climate is typically monsoonic with an annual rainfall of 1051 mm in 1989, about 96 per cent of which occurred between June to September. The mean monthly maximum and minimum temperatures during 1989 ranged from 20.7 to 39.2°C and from 6.6 to 26.3°C, respectively.

A total of one experimental site each with 6, 8, 12, 15 and 25 year old mango (*M. indica* var. Dashehari) trees was selected for the present study. Thus, a total of 5 sites were selected. These sites are located in close proximity with each other. Further, to minimize intersite variations, sites were chosen with similar soil type (i. e. sandy loam). The distance among the trees was 8 m on all the sites except on site with 25 year old trees where the distance was 10 m. On all the sites wheat (*Triticum aestivum* L. var. HD 2285) was grown in the interspaces of mango trees as per its recommended cultural practices. The crop was given sufficient fertilizer and irrigation.

Five individuals of mango from each site were chosen at random for the measurement of tree height, canopy depth, canopy width and stem diameter (at 30 cm from the ground). Height of trees and canopy depth were measured by a hypsometer (Forbes, 1961). Three, 30 m line transects were established in each site and the lengths of the interception by the canopy of trees was recorded. This was then calculated as percentage of the total length of the transect to provide an estimate of canopy cover.

For the estimation of growth and yield of wheat, eight quadrats of 50 × 50 cm size were placed around each individual tree in the last week of April, 1990 when the crop was ready for harvesting. Four of these quadrats were placed under the tree canopy and four outside the canopy (i. e. in open area). Thus, the data on growth and yield of wheat on each site are based on 20 quadrats each under canopies and in open areas. To eliminate bias, quadrats were consistently placed at each direction of the compass. All the plants of wheat rooted in the quadrat were harvested at ground level and were brought to the laboratory for the estimation of total number of plants, plant height, number of tillers, panicle length, number of grains and yield of grains and by-products.

Light intensity was measured at each quadrat and averages were computed for the under canopy and open area quadrats associated with trees of different sites. The data were collected at the time of crop harvesting at 10:00 a. m., 12:00 noon and 2:00 p. m. by using Lux meter.

RESULTS AND DISCUSSION

Tree Characteristics

Characteristics of mango trees on various sites are summarized in Table 19.1. The average height of mango trees ranged from 2.2 m on site 1 to 11.2 m on site 5. On an average, about 77 to 88 per cent length of the trees was covered by the canopy. Further, the canopies on all the sites were wider than deeper. The canopy cover was minimum on youngest site (site 1), while on oldest site (site 5) the canopy is well developed and it overlapped with the canopies of adjacent trees. The annual increment in stem diameter was about 1.3 cm up to 15 years. However, between 15 to 25 years, the diameter increment of mango trees was about 2 cm per year.

Table 19.1: Characteristics of mango trees on various sites (values are averaged for each site)

Parameter	Site 1	2	3	4	5
Age of the trees (yr)	6.0	8.0	12.0	15.0	25.0
Tree height (m)	2.2	2.7	3.5	4.9	11.2
Canopy depth (m)	1.7	2.1	2.9	4.3	9.5
Canopy width (m)	1.9	3.1	3.9	4.5	11.8
Stem diameter (cm)	8.6	10.2	17.5	19.4	49,7
Canopy cover (%)	23.8	38.8	48.7	56.2	100.0

Light Intensity

As expected, light intensity values were lower under the tree canopies than in open areas. The data on light intensity under tree canopies at different sites are given in Table 19.2. The average light intensity under the canopies ranged from 6 per cent light in open on site 5 to 20.7 per cent on site 1. Thus, 94 per cent light was the canopies of 25 year old mango trees and this reduction was progressively decreased as the age of trees decreased reaching to a minimum reduction of about 79 per cent under 6 year old trees.

Table 19.2 : Relative light intensity (% of sun light in open) under tree canopy on different sites

Site	Time: 10 : 00 a. m	12 : 00 noon	14 : 00 p.m.	Average
1.	22	20	20	20.7
2.	18	12	17	15.7
3.	13	9	12	11.3
4.	8	7	8	7.7
5.	7	6	5	6.0

Light intensity under canopies decreased with increasing canopy depth and tree height (Fig. 34.1). However, the decrease was rapid up to the canopy depth of about 4 m and tree height about 5 m and thereafter the reduction in light intensity was slow.

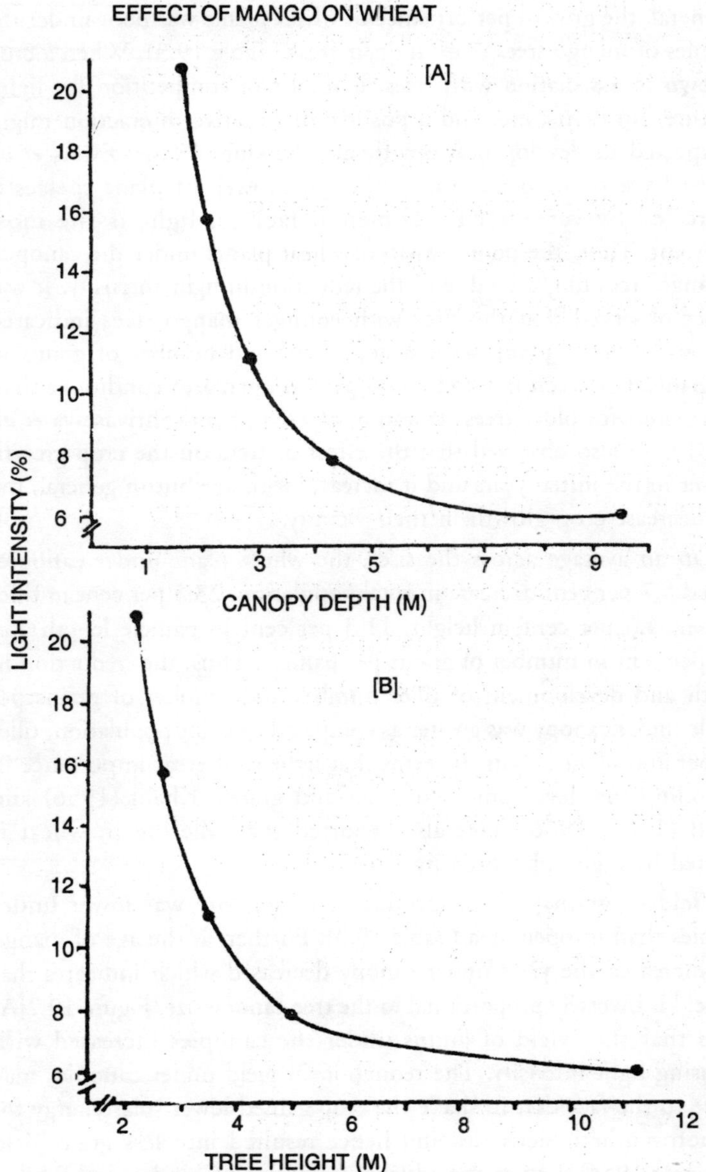

Fig. 19.1 : Effect of canopy depth [A] and tree height [B] on light intensity (% of light in open) under tree canopies.

Growth and Yield of Wheat

In general, the growth performance of wheat plants was poor under the canopies of mango trees than in open areas (Table 19.3). When a crop is grown in association with trees, it results in competition for light, moisture, nutrients, etc. and a positive or negative interaction might be expected to develop between them (Hocking, 1986). Evans *et al.* (1976) have pointed out that although growth of plant species is controlled by various environmental factors, light is the most important. Thus, the poor growth of wheat plants under the canopies of mango trees might be due to the reduction in light intensity. It was further observed that the sites with younger mango trees indicated lower variation in plant population, height and number of grains in each panicle between the under canopy and open area conditions than on the site with older trees. Prasad *et al.* (1983) and Shrivastava *et al.* (1983) have also observed that the effect of trees on the crop growth is lower in the initial years and it increases with age but in general, the trees decrease crop growth in their vicinity.

On an average across the sites, the wheat plant under canopies showed 5.7 per cent decrease in total population, 23.5 per cent in tiller number, 4.6 per cent in height, 11.3 per cent in panicle length and 21.6 per cent in number of grains per panicle. Thus, the reduction in growth and development of tiller number and number of grains per panicle under canopy was greater as compared to plant population, tiller number and plant height. It seems that light is of great importance in controlling the development of tiller and grains. Khalil (1956) and Friend (1965, 1966) have also reported that tillering in wheat is favoured by high light intensity.

Yield of grains and by-products at each site was lower under canopies than in open area (Table 19.3). Further, as the age of mango trees increased the yield under canopy decreased which indicates that the yield is inversely proportional to the tree canopy size. Figure 19.2[A] shows that the yield of grains under the canopies increased with increasing light intensity. The reduction in yield under canopies may be due to the fact that in shade the crop gained lower solar energy for its photosynthetic activities and hence resulted into less grain yield (Wassinck, 1954). Longman and Jenik (1974) and Chazdon and Fetcher (1984) have opined that the light regime on tropical forest floor is

Table 19.3 : Growth parameters and yield of wheat under tree canopy and in open area on various sites

Parameter	*Site*											
	1		*2*		*3*		*4*		*5*		*Average*	
	Under canopy	*Open*	*Under canopy*	*Open*	*Under canopy*	*Open*	*Under canopy*	*Open*	*Under canopy*	*Open*	*Under canopy*	*Open*
Plant population ('000/ha)	872	990	1162	1033	839	883	1001	1022	441	893	887	941
Tiller number (per plant)	1.7	2.3	1.4	1.9	1.4	1.7	1.2	1.4	1.0	1.2	1.3	1.7
Plant height (cm)	77.4	73.5	70.1	73.2	69.5	72.1	68.6	71.7	57.6	69.2	68.6	71.9
Panicle length (cm)	7.2	8.3	7.2	8.2	7.4	8.1	6.9	7.9	6.3	7.6	7.1	8.0
Number of grains (per panicle)	22.6	25.9	23.2	25.5	23.9	25.6	20.6	27.3	13.2	27.5	20.7	26.4
Grain yield (q/ha)	15.2	17.8	11.5	17.6	8.5	15.0	7.4	14.8	3.3	8.5	9.2	14.7
Yield of by-products (q/ha)	16.9	20.6	15.7	19.5	11.0	17.2	11.1	18.2	4.3	10.2	11.8	17.1

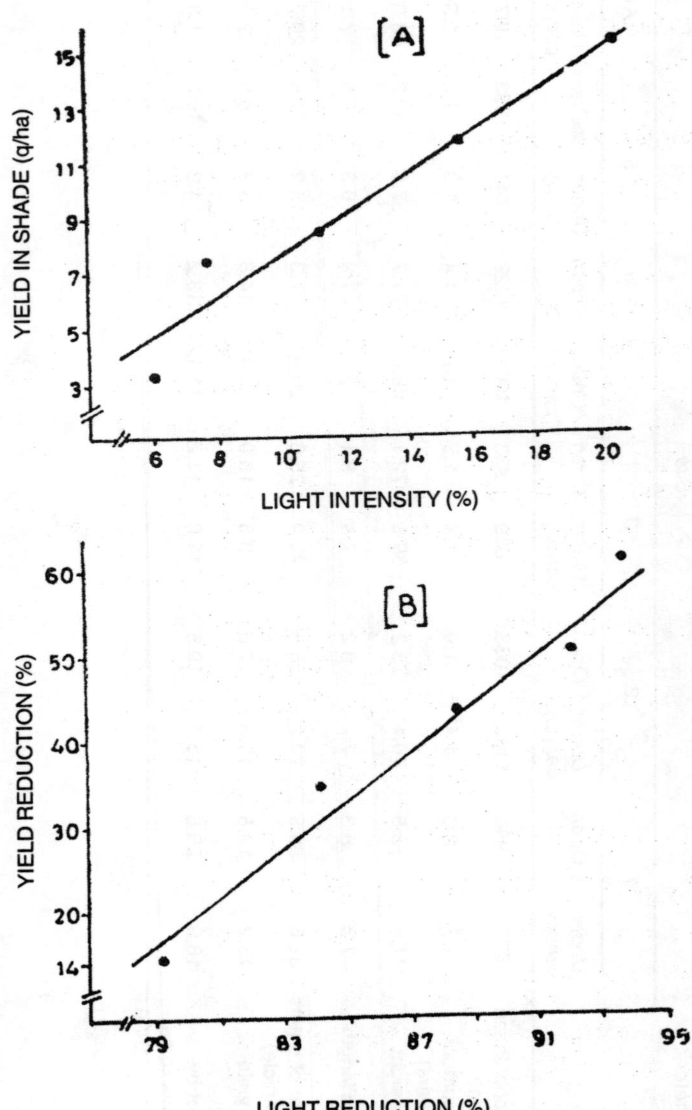

Fig. 19.2 : Relationship between : (A) grain yield and light intensity, and (B) reduction in grain yield and light intensity under tree canopies.

characterized by a very low phyton flux density and a low red/far red ratio which might be important in the development of photosynthetic capacity of plants. It seems that when the supply of assimilate from the leaves is reduced by shading, grain yield is also reduced (Bremner, 1972).

On an average across the sites, the grain yield was 14.7 q/ha in open and 9.2 q/ha under canopy indicating a reduction of about 38 per cent in grain yield under canopy. Similarly, the average yield of by-products was also reduced by about 31 per cent under mango tree.

The values for reduction in grain yield of wheat recorded under the canopies of certain other trees and for the present study are given in Table 19.4. The values for the present study are lower than those reported for *Eucalyptus* sp. and *Acacia nilotica* by Dhillon *et al.* (1982, 1984).

Table 19.4 : Per cent reduction in grain yield of wheat under various tree species

Tree species	*Age (yr)*	*Reduction in yield(%)*	*Source*
Mangifera indica	6	14.5	Present study
M. indica	8	34.7	—do—
M. indica	12	43.2	—do—
M. indica	15	50.0	—do—
M. indica	25	60.7	—do—
Eucalyptus sp.	–	64.4	Dhillon *et al.* (1982)
Acacia nilotica	9-11	65.3	Dhillon *et al.* (1984)
Dalbergia sissoo	10-15	47.4	—do—

On the other hand, the reduction in grain yield is almost equal under 12-15 year old mango trees and 10-15 year old *Dalbergia sissoo* trees. Allelopathy from *Eucalyptus* sp. and *A. nilotica* could be a major factor associated with greater reduction of yield under the canopies of these species (del Moral and Muller, 1969; Dhillon, *et al.* 1984). In the present study, the reduction in grain yield under canopy, which ranged from 14.5 to 60. per cent, shows a positive relation with the reduction in light intensity (Fig. 19.2 [B]).

When the data on grain yield are averaged across the locations (i.e. under canopy and open area), the grain yield on the site with 25 year old mango trees was only 5.9 q/ha as compared to 16.5 q/ha on the site with 6 year old mango trees. The average yield shows a negative relation with canopy cover of mango trees (Fig. 19.3) which again reflects the importance of light on the grain yield of wheat.

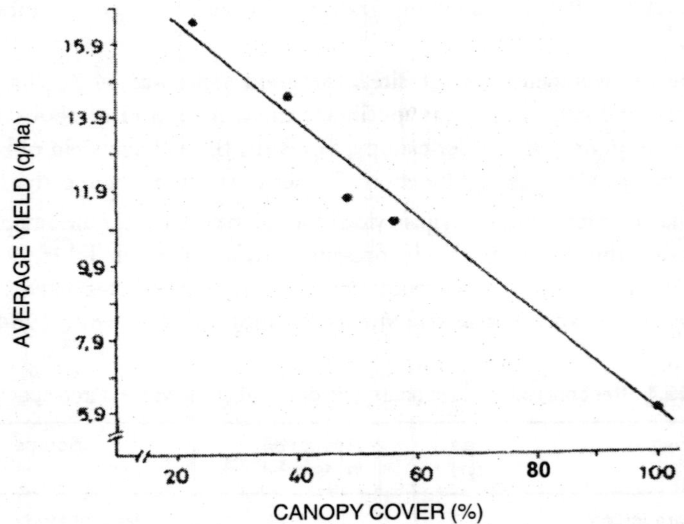

Fig. 19.3 : Relationship between average grain yield and tree canopy cover.

REFERENCES

Anonymous (1983). *Mango cultivation.* Indian Institute of Horticultural Research, Bangalore, p. 25.

Bremner, P. M. (1972). "The accumulation of dry matter and nitrogen by grains in different positions of the wheat ear as influenced by shading and defoliation." *Aust. J. Biol.* Sci. 25 : 653-681.

Champion, H. G. and Seth, S. K. (1968), *A revised survey of the forest types of India.* Govt. of India Publ., Delhi, p. 404.

Chazdon, R. L. and Fetcher, N. (1984). "Light environments of tropical forests." In F. Medina, H. A. Mooney and C. Vazques-Yanes (eds.) *Physiological ecology of plants in the wet tropics.* Junk, Holland, pp. 27-36.

del Moral, R. and Muller, C. H. (1969). "Fog drip : A mechanism of toxin transport from *Eucalyptus globulus.*" *Bull. Torrey Bot. Club* 96 : 467-475.

Dhillon, G. S., Singh, S., Dhillon, M. S. and Atwal, A. S. (1982). "Developing agri-silvicultural practices : studies on the shading effect of *Eucalyptus* on the yield of adjoining crops." *Indian J. Ecol.* 9 : 228-236.

Dhillon, M. S., Singh, S., Atwal, A. S, and Dhillon, G. S. (1984). "Developing agri-silvicultural practices : effects of *Dalbergia sissoo and Acacia nilotica* on the yield of adjoining crops." *Indian J. Ecol.* 11 : 249-253.

Evans, G. C., Bainbridge, R. and Rackhan, O. (1976). *Light as an ecological factor.* Blackwells, Oxford.

Forbes, R. D. (1961). *Forestry handbook.* Ronald Press, New York.

Friend, D. J. C. (1965). "Tillering and leaf production in wheat as affected by temperature and light intensity." *Canad. J. Bot.* 43 :1063-1076.

Friend, D. J. C. (1956). "The effects of light and temperature on the growth of cereals." In F. L. Milthorpe and J. D. Ivins (eds.) *The growth of cereals and grasses.* Butterworths, London. pp. 181-199.

Hocking, D. (1986). "Some principles of crop complementarity in agroforestry systems." In P. K. Khosla, S. Puri and D. K. Khurana (eds.) *Agroforestry systems — A new challenge.* ISTS, Solan, pp. 39-42.

Khalil, M. S. H. (1956). "The interrelation between growth and development of wheat as influenced by temperature, light and nitrogen." *Meded. Landbouwhogesch. Wageningen* 56 : 1-73.

Longman, K. A. and Jenik, J. (1974). *Tropical forest and its environment.* Longman, New York.

Prasad, A., Subbayan, R., Sam, M. and Ramnath (1983). "Preliminary studies on the effect of different tree species on yield of field crops under rainfed condition in Vertisols of Ballary." Presented in *Nat. symp. on watershed management for increased productivity in red and black soil.* Mimeo, p. 6.

Shrivastava, A. K., Verma, B. and Narain, P. (1983). "A case study for farm forestry in South-eastern Rajasthan." Presented in *Nat. symp. on watershed management for increasing productivity in red and black soils.* Mimeo, p. 4.

Wassinck, E. C. (1954). "Remarks on energy relations in photosynthesis processes." In *Proc. lst Int. Photobiol. Congr., Amsterdam. Biology Sect. V.*

20

Scope of Power Tiller for Agroforestry Management

M.R. Varma and *P. M. Singh*

At present, almost every operation for raising agroforestry is being done manually except transport. The canopy of the trees restricts the use of four wheel tractors for alley management and other operations in agroforestry. However, the operations like seed-bed preparation for nursery raising, pit digging, water pumping, interculture, alley management, plant protection, transport, etc. could be done easily and efficiently using power tiller (two-wheel tractor) with their matching equipment. Moreover, the power tiller can be easily manoeuvred on undulated, degraded and small piece of land generally utilized for agroforestry. It has also been found that use of power tiller for various operations is economical in comparison to bullock and tractor power systems.

Key Words : Power tiller, alley management, agroforestry.

The power tiller, a two-wheel tractor, with matching equipments is being used successfully for agriculture farming in various parts of the country. Although, it can also be used as a potential source of power for the management of agroforestry, there is need of proper demonstration for its popularization in doing various operations related to agroforestry. In the present chapter, the scope of power tiller for different operations in the management of agroforestry is discussed.

Nursery Management

Seedlings are grown mostly in polypots filled with pulverized soil and Farm Yard Manure (FYM) and sometimes directly in the well prepared seed-beds. On large scale, small grinder operated by a power tiller engine may be used for pulverizing the soil and FYM. A good quality nursery bed may also be prepared by single ploughing and twice rotatilling using a power tiller. The power tiller can also be used in small plots as its turning radius is small. Table 20.1 shows that the cost of seedbed preparation with manual labour is about Rs. 1854.50/ha, whereas with power tiller it costs only about Rs. 888.38/ha. Thus, there is a saving of about 52 per cent in the cost of nursery seed-bed preparation. Similarly, manual seed-bed preparation required about 22 times much labour than in the seed-bed preparation through power tiller system.

Field Layout and Pit Digging

At present, field layout for marking actual position of pits is done manually using ropes and pegs. These markings could be done easily and quickly by using marker with power tiller. Similarly, for planting the young seedlings pits are dug manually which is also labour intensive and involves drudgery. This operation can be done easily by post hole digger operated by power tiller. The commercially available post hole diggers can dig holes of 30 × 40 cm size in one pass and of greater sizes in more than one passes.

The cost of manual digging of 30 × 40 cm size pit is about Rs. 0.75 to 1.00 per pit, whereas with power tiller post hole digger it costs about Rs. 0.70 per pit (Anonymous, 1987). Further, the studies conducted by Abrol and Gill (1984) indicated that auger holes of 10 cm × 120 cm as well as 15 cm × 120 cm gave 100 per cent survival and better girth growth rate of *Eucalyptus tereticornis* and *Acacia nilotica* in highly alkaline soil. The study also indicated that auger hole technique is economical and less troublesome than the pit method generally recommended for alkali soil.

Irrigation of Nursery and Plants

The irrigation of plants raised on degraded and marginal soil is difficult due to the undulated topography of land. Hence, irrigation through light weight portable gated pipes is a suitable method of irrigation for these situations. The power tiller with pump may be used successfully for this purpose as it could be manoeuvred and transported from place to place easily and quickly.

Table 20.1 : Comparative cost of nursery seed-bed preparation by manual v/s power tiller

Manual			*Power tiller*			
Operation	*Labour requirement (m-hr/ha)*	*Cost (Rs./ha)*	*Operation*	*Machine (mc-hr/ha)*	*Labour (m-hr/ha)*	*Cost (Rs./ha)*
Ist Digging by spade	450	647.00	Ploughing	16.7	16.7	332.20
Clod breaking & grass picking	200	287.50	Ist Rotatilling	11.1	11.1	234.59
IInd Digging	320	460.00	IInd Rotatilling & mixing of FYM	11.1	11.1	234.59
FYM mixing-cum digging	320	460.00	Planking	4.0	4.0	64.00
			Preparation of corners (manual)		16.0	23.00
Total	1290	1854.50		42.9	58.9	888.38

Labour wages @ Rs. 11.50/day.

Interculture

Weeding and hoeing of plantation crop is essential for their proper growth. A power tiller with suitable equipment may be used for these operations very efficiently. Table 20.2 shows that interculture cost with power tiller by single ploughing, rotatilling and planking plus utilizing human labour for digging just beneath and around the plants is Rs. 708.08/ha, whereas by using manual labourers it costs Rs. 934.50/ha for single digging, clod breaking and grass picking. Thus, there is a cost saving of 24.23 per cent and at the same time there is labour saving of about 70 per cent.

Plant Protection

The experiment carried out under power tiller scheme at different

Table 20.2 : Comparative cost of interculture by manual v/s power tiller

Manual			*Power tiller (8-10 hp)*			
Operation	*Labour requirement (m-hr/ha)*	*Cost (Rs./ha)*	*Operation*	*Machine requt (mc-hr/ha)*	*Labour requt (m-hr/ha)*	*Cost (Rs./ha)*
Digging by Kudal	320	647.00	Ploughing	13.5	13.5	270.00
Clod breaking and grass picking	200	287.54	Ist Rotatilling	9.0	9.0	190.00
			Planking	4.0	4.0	64.00
			Digging around plants (manually)	–	128.0	184.00
Total	520	934.50		26.5	154.5	708.00

research centres showed that spraying by power tiller with ASPEE-HTP power sprayer is 30-40 per cent cheaper than manual spraying (Anonymous, 1986).

Alley Management

The power tillers with matching equipment are the appropriate and better source of power for a alley management too. Grass cutting in allies could also be done easily by using power tiller operated grass cutter. The cost of grass cutting by power tiller including collection was about Rs. 244/ha, whereas with manual labour it was estimated at Rs. 560/ha.

Harvesting

In India, harvesting of trees is done mostly manually by cutting with an axe. However, this operation could easily be done by using portable power tiller operated circular or chainsaw. Efforts are being made to design and fabricate this new device at Central Institute of Agricultural Engineering, Bhopal.

Transportation

The transportation of fuel wood collected from forests is generally done as head load. In some cases, bullock carts and other transport devices like tractors and trucks are also used. The power tiller with its trailer can also be used for this purpose. The capacity of power tiller trailer is about 1 tonne and it can be pulled at a speed of 15 km/hr on tarmacadam roads.

It is thus concluded that power tiller with its matching equipment for different operations is an appropriate and suitable source of power for agroforestry and orchard management as well. The requirement of equipment for agroforestry systems are quite different and location specific and hence no general recommendation of package of tools and equipment can be given.

REFERENCES

Anonymous (1986). *Annual Report (MPAU Rahuri Centre) of ICAR all-India coordinated project on power tiller, etc.* Fourth annual workshop at VST Tillers Tractors Ltd., Bangalore, Jan. 28-30, 1987, pp. 20-33.

Anonymous (1987). *Annual Report (Bhopal Centre) of ICAR all-India coordinated project on power tiller, etc.* Fifth annual workshop at OUAT, Bhuvaneshwar, Jan. 8-10, 1988, pp. 26-30.

Abrol, I. P. and Gill, H. S. (1984). "Afforestation of salt-affected soils." *Proceedings of bio-energy society first convention* and *symposium*, 14-16 October, 1984, New Delhi, pp. 37-47.

Index